ScienceWorld 8

Australian Curriculum edition

Lesley Englert · Peter Stannard · Ken Williamson

BA, DipT · BSc, DipEd · BSc (Hons), DipEd

This edition published in 2021 by

Matilda Education Australia, an imprint
of Meanwhile Education Pty Ltd
Melbourne, Australia
T: 1300 277 235
E: customersupport@matildaed.com.au
www.matildaeducation.com.au

First edition published in 2011 by Macmillan Science and Education Australia Pty Ltd

Publication data

Author: Lesley Englert, Peter Stannard and Ken Williamson

Title: ScienceWorld 8 Workbook: Australian Curriculum edition
ISBN: 978 1 4202 2995 0

Publisher: Peter Saffin
Project editor: Claire Lavin
Illustrator: Chris Dent, and Andy and Nives Porcellato
Cover designer: Dim Frangoulis
Text designer: Firefly Education
Production control: Loran McDougall
Typeset in Helvetica, Trade Gothic and Linotype Kaliber by Firefly Education
Cover image: Photolibrary/Volker Steger (main photo), Shutterstock/Annett (background detail)

Printed in China by Central
Sep-2024

About this book

This workbook is full of exercises which will help you understand the material in the textbook, think about the information, relate it to your own experiences, and write about it in your own words.

The exercises ask questions and pose problems to give you the opportunity to reflect on your knowledge, then show what you know.

Do you find that you know the answer but often do not know how to write it in the correct form? The workbook teaches you how to organise this information into many different written and oral forms.

On the following page you will see some sample pages from the workbook. You will also see descriptions of the various features in the workbook. These descriptions will help you become familiar with the workbook.

Getting to know your workbook

Each chapter begins with the same chapter number and name as found in your textbook.

Sometimes you will see a Challenge box. These boxes contain challenging questions or problems that you might like to attempt.

Chapter 10

Rocks

Overview

Igneous rocks
Volcanic
Plutonic
Sedimentary rocks
Metamorphic rocks
Magma
cooling and hardening
weathering and erosion and deposition
heat and pressure
heat and pressure
melting

What do you know already?

Exercise 1 Discussing in groups

Discuss the following topics in small groups.

1 What is the inside of the Earth like?

2 What causes volcanoes?

3 How was the Grose Canyon formed? (See Fig 11 on page 235.)

4 How are rocks changed into soil?

5 Where do rocks come from? Why are there many different types of rocks? Can you name at least three different types?

6 What are rocks made of?

74 ScienceWorld 8 Australian Curriculum edition Workbook 978 1 4202 2995 0

This is called the Overview—it is a diagram of the way the ideas in the chapter are connected.

The section of the workbook called During reading contains exercises which ask you questions about the ideas on certain pages of your textbook.

The first section of the workbook is called What do you know already? This will show how much you already know about the work in the chapter. This is not a test.

During reading

Exercise 2 Summarising

After reading Properties of matter on page 100

In the Activity on page 101 you used a table to summarise the properties of solids, liquids and gases.

1 Complete this sentence comparing and contrasting solids and liquids.

Solids and liquids have similar ____________, but liquids do not have a ____________.

2 Now it's your turn. Write a sentence comparing and contrasting liquids and gases. Use *comparison and contrast linking words* from page 92 of this workbook.

Exercise 3 Clarifying ideas and concepts

After reading Density on pages 102–103

Complete the cartoons below.

Chapter 5 – Particles 978 1 4202 2995 0 35

The exercises contain questions for you to answer. Sometimes they require you to write words in the space provided. Others ask you to draw data tables or graphs. Some ask you to create stories or to solve problems.

Chapter Roundup will check how much you know about the main ideas in the chapter. Be prepared to write answers, teach someone else, have a group discussion, or role play.

Chapter 1
Science at work

Overview

Science is a way of finding answers to questions by using experiments

Observing

Controlling variables

Collecting data (tables and graphs)

inferences | Interpreting data | generalisations

predictions | Writing reports | hypotheses

What do you know already?

Exercise 1 Investigating

Ken, Peter and Maria want to find out which of them is the strongest, so they decide to conduct an investigation using weights.

Ken uses a lever to lift 80 kg. Peter lifts 50 kg from a lying position. Maria lifts 30 kg from a standing position.

1 Does this mean that Ken is the strongest? Explain your answer.

2 Describe how you would do this investigation so that it is a *fair test* (an experiment where you change something, measure something, and keep everything else the same).

During reading

Exercise 2 Matching words with meanings

After reading page 3

Without looking at your textbook, match each word to its meaning.

generalising	• trying to explain your observations
observing	• examining, inquiring into
predicting	• a well thought out test
recording	• writing down your data
inferring	• making a forecast of what a future observation will be
investigating	• observations described with measurements
data	• observations described in words
qualitative data	• writing a statement that seems true in most cases
quantitative data	• information gathered
experiment	• using your senses to find out about an object or event

Exercise 3 Matching illustrations and captions

After reading page 3

Select the appropriate caption (the word in **bold** type) from the list and write it below its illustration.

CAPTIONS • **Observing** • **Recording** • **Inferring** • **Predicting** • **Generalising**

____

____

____

____

____

Exercise 4 Writing contrast sentences (to explain differences)

After reading page 3

Complete the sentences. Note that the contrast linking words are in *italics*.

1 Observing is using your senses to find out about an event, *whereas* recording is

____

2 Inferring is trying to explain your observation, *while* predicting is

____

Exercise 5 Analysing text

After reading Investigations and experiments on page 4

Here is a paragraph from the textbook. Read it carefully then answer the questions below.

What's the difference between an investigation and an experiment? The terms mean much the same thing. Both involve carefully planned laboratory or field work. However, an experiment is based on solving a problem or answering a question.

1 Find a comparison linking word.
(This word signals that the information following will describe a similarity between the two terms.)

____

____

2 Find a contrast linking word.
(This word signals that the information following will describe a difference from the information before it.)

____

____

3 In one sentence compare and contrast the words investigation and experiment. Use the linking words *similar* and *on the other hand*.
An experiment ____

____

____

Exercise 6 Identifying and controlling variables

After reading Controlling variables on page 7

Anthony plants beans and plays rock music to them for 3 hours a day. The beans grow 15 cm in 2 weeks.

Debbie plants a cactus and plays rock music to it for 10 hours a day. It grows 1 cm in 6 months.

Jessica plants roses and plays Christmas carols to them for 2 hours a day. The roses die within a month.

1 Does this mean that roses hate Christmas carols or that beans like rock music more than a cactus does? ______

Why or why not? ______

2 If you want to find out if plants respond to music:

a What must you change? ______

b What must you keep the same? ______

3 Why is it a good idea to keep at least one plant which does not have music played near it?

______

______

______

4 Draw the set-up needed to show that plants respond to music. Add captions to the set-up.

Exercise 7 Hypothesising (or writing a hypothesis)

After reading page 8

1 In this chapter, you learnt about generalisations. A hypothesis is a generalisation which can be tested. Tick the generalisations below which you could test in an experiment. (Generalisation linking words are in *italics*). In a group, discuss how you would test each hypothesis.

a Heavy people go faster down a waterslide.

b Cats are *usually* more intelligent than dogs.

c Women are *generally* better drivers than men.

2 Complete the table below.

Observations	David can lift heavier weights than Alison. Carlos can lift heavier weights than Beth. Stefan can lift heavier weights than Mary-Louise. Mr Bloggs can lift heavier weights than Ms Smales.
Generalisation (Hypothesis that can be tested)	
Prediction	

Exercise 8 Recording in data tables

The best way to record your observations or measurements in a form that is easy to understand is to use a data table. Here are some data tables.

My age	My height
2 years	0.9 m
3 years	1.1 m
4 years	1.3 m
8 years	1.48 m
10 years	1.55 m

Planet	Diameter (Compared to Earth = 1)	Mass	Gravity
Mercury	0.4	0.06	0.4
Venus	0.9	0.8	0.9
Jupiter	11.2	318	2.6
Saturn	9.4	95	1.1

A cyclist decides to collect data on the relationship between the speed at which he travels and his pulse rate. He is careful to control variables, and records his measurements in a table, shown on the right.

Cyclist's results

Speed (km/h)	Pulse rate (beats/min)
0	70
10	95
20	120
30	142
40	165

Use the data table *Cyclist's results* to answer these questions.

1 What is his pulse rate when his speed is 10 km/h? ____________

2 What is his speed when his pulse rate is 142 beats/min? ____________

3 Predict what his speed will be when his pulse rate is 100 beats/min ____________

4 Predict what his pulse rate will be when his speed is 50 km/h ____________

5 Complete the following generalisations linking the cyclist's speed and his pulse rate.

As the cyclist increased his speed, his ____________________________________.

OR The faster the cyclist travelled, the ____________________________________.

Exercise 9 Creating and interpreting graphs

After reading Graphing on page 10 and the Skillbuilder on pages 10–11

A graph is an excellent way of showing the relationship between two measurements. Here are some graphs.

Bar graph

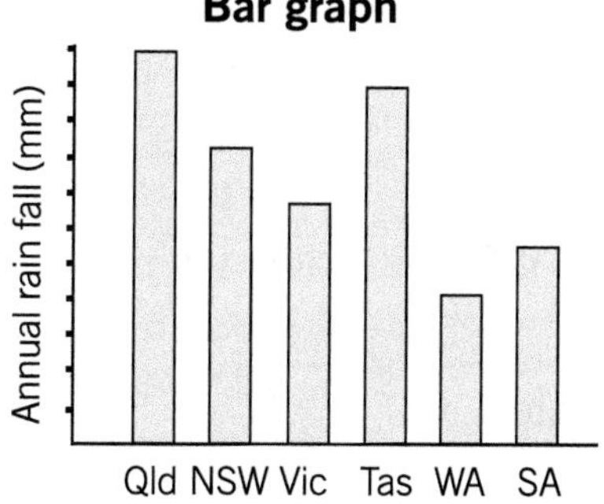

Line graph

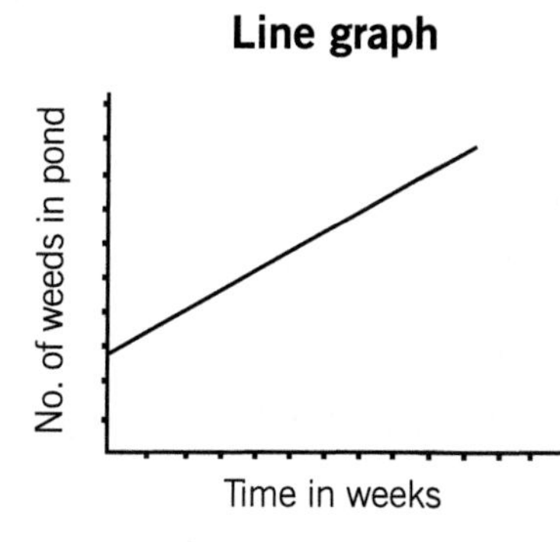

Pie chart

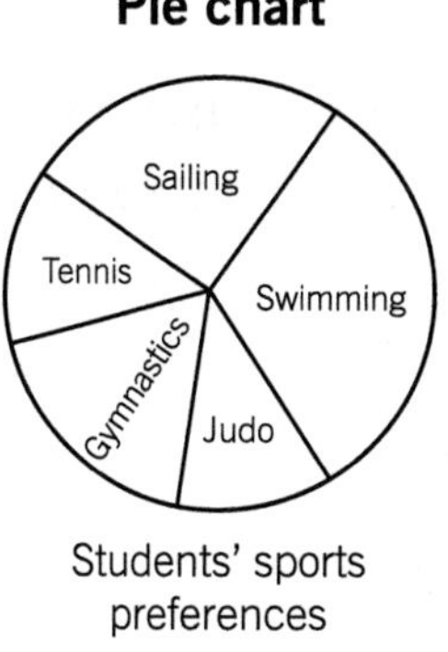

Students' sports preferences

1 Look at the bar graph above.

a What is the independent variable (on the horizontal axis)? ______________________

b What is the dependent variable (on the vertical axis)? ______________________

2 Look at the line graph above.

a Which is the independent variable? ______________________

b Which is the dependent variable? ______________________

c Complete these generalisations:

As time goes by, the number of weeds in the pond ______________________.

OR The longer the time, the ______________________.

3 Look at the pie chart above.

a Which sport is the most preferred?

____

b Which sport is the least preferred?

____

c What is the dependent variable?

____

4 Use the grid on the right to draw a line graph using the data from *Cyclist's results* in exercise 8 on the previous page.

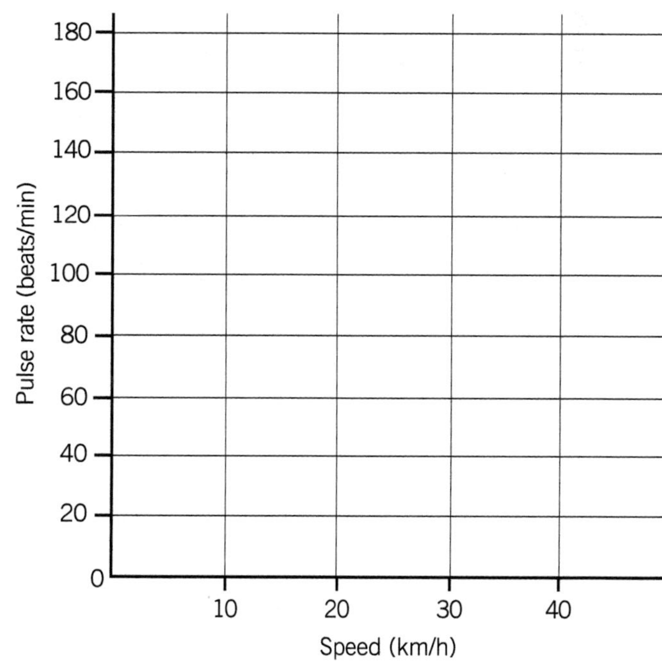

5 From your graph:

a Predict what his pulse rate will be when the speed is 5 km/h.

____

b Predict what the speed will be when his pulse rate is 130 beats/min.

____

Exercise 10 Writing a paragraph

A paragraph is a 'parcel' of information, usually containing several sentences which describe or develop a main idea. Paragraphs usually begin with a topic sentence which states the main idea of the paragraph. (This main idea is often a generalisation.) The remaining sentences in the paragraph then give details supporting the topic sentence.

This is the structure for a paragraph.

TOPIC SENTENCE + SENTENCES WITH SUPPORTING DETAILS

Using this structure, you can write a paragraph describing the data in *Cyclist's results* in exercise 8, and the graph in exercise 9.

Topic sentence (your generalisation) → The faster the cyclist travelled, the higher his pulse rate.

Sentences which support the topic sentence → At rest, his pulse rate was 70 beats/min. However, when he cycled at 10 km/h his pulse rate reached 95 beats/min. As his speed continued to increase, so did his pulse rate. When he was moving at 40 km/h, his pulse rate increased to 165 beats/min.

Using the paragraph above as a model, write a paragraph interpreting the results in the bar graph on the right.

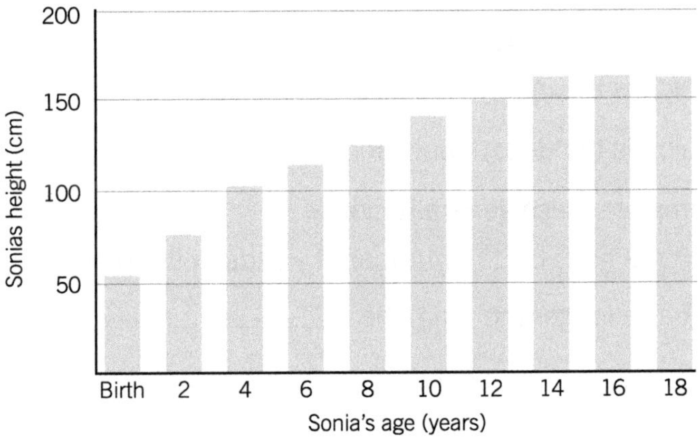

Topic sentence (your generalisation):

____

____

____

____

Sentences which support the topic sentence:

____

____

____

____

____

Roundup

Exercise 11 Introducing and thanking a guest speaker

After reading Science as a Human Endeavour on pages 19–21

Imagine that either Dr Helen Newton Turner or Dr Michael Tyler has agreed to visit your class as a guest speaker. Guest speakers should be introduced before they begin, and thanked when they finish.

If you have the task of introducing a guest speaker you should prepare by following these steps.

- Find out some background information about the speaker—by researching in books, magazines, newspapers or on the internet, by interviewing the speaker's friends, or by interviewing the speaker to find out some interesting facts.
- Think about the audience by asking yourself:
  'If I was in the audience, what would interest me about this speaker?'
  'Which information about this speaker will grab the listeners' attention and make them want to listen carefully to the presentation?'
- Write your introduction, keeping it as interesting as possible, and then re-write it onto a palm card, with only the headings and main points listed. (You will then be able to focus on your audience, rather than keeping your eyes down, reading.)
- Practise your introduction several times beforehand, preferably to an audience—to your family or friends, in the mirror, or even to your dog!

An introduction usually begins like this: 'Teachers and students, it is my pleasure to introduce our guest speaker, Dr Helen Newton Turner/Dr Michael Tyler, who will be speaking to us today about her/his research in

____

____

and ends with: 'Please join me in welcoming Dr Helen Newton Turner/Dr Michael Tyler.'
(You start clapping and the audience joins in.)

Using the information on pages 19–21, prepare an introduction for Dr Helen Newton Turner or Dr Michael Tyler.

A member of the audience should be organised beforehand to give a vote of thanks to the guest speaker. If you have this task, you should prepare by

- knowing the general structure of a vote of thanks (below); and
- taking notes during the presentation on facts that you found interesting.

A vote of thanks usually begins like this: 'On behalf of the students and of XYZ class teachers, I would like to thank Dr Helen Newton Turner/Dr Michael Tyler for her/his very interesting and informative presentation. I was really interested in

____

____

and ends with: 'Please join me in thanking Dr Helen Newton Turner/Dr Michael Tyler.'
(You start clapping and the audience joins in.)

Imagine that Dr Helen Newton Turner or Dr Michael Tyler mentioned all the facts on pages 19–21 in the presentation. Prepare a vote of thanks for her/him.

Chapter 2

Chemical reactions

Overview

CHANGES

Physical changes
- No new substances formed
- Properties of original substances may change
- Usually reversible

Chemical reactions
- New substances formed (products)
- New substances have different properties
- Not easily reversible

Evidence of chemical reaction may be
- gas
- colour change
- precipitate
- heat
- light
- sound
- electricity

Common gases
- oxygen
- hydrogen
- carbon dioxide

What do you know already?

Exercise 1 Discussing in groups

In small groups, discuss and decide on answers for the following questions.

1 When you tell a joke, you get a reaction. What is a reaction?

____

____

2 What do you think a chemical reaction is?

____

____

3 Complete the table below. The first is done for you as an example. Signs that a chemical reaction have taken place are: colour change, gas produced, heat produced, light produced, sound produced, elements produced.

Common example	Signs that a chemical reaction has taken place
Sparkler burning	Light, heat, sound, gases, colour change.
A motor turning when connected to a battery	
Dynamite exploding	
Iron roof rusting	
Vinegar added to baking soda	
Coal burning	

During reading

Exercise 2 Making and justifying decisions

After reading What is a chemical reaction? on pages 31–32

Read the statements below and decide whether each one is true or false. Circle your choice and give a reason for your decision.

1 The properties of ice are the same as the properties of water. TRUE FALSE

____

____

2 The change that occurs when water freezes can be reversed. TRUE FALSE

____

____

3 When you turn on a light, a chemical reaction occurs. TRUE FALSE

____

____

4 Heat and light are sometimes produced during a physical change. TRUE FALSE

____

____

5 New substances are produced in a bushfire. TRUE FALSE

____

____

6 Some chemical reactions produce heat and light. TRUE FALSE

____

____

Exercise 3 Completing cartoons

After doing Investigation 4 on pages 32–34 and reading Reactants and products on page 35

Complete the following cartoons.

Investigation 4 PART A Reaction 1
dilute hydrochloric acid and magnesium

Investigation 4 PART A Reaction 2
copper sulfate and magnesium

Investigation 4 PART A Reaction 4
copper sulfate and sodium hydroxide

Investigation 4 PART A Reaction 7
potassium permanganate and hydrogen peroxide

Investigation 4 PART B
copper and magnesium, aluminium or steel in a lemon

A chemical reaction occurred because

Exercise 4 Labelling

After reading Reactants and products on page 35

Here are three chemical equations. Complete each equation and, on the line under each substance, label it as a **reactant** or a **product**.

1	baking soda	+ ______	→	carbon dioxide
	______	______		______
2	wood	+ oxygen	→	______
	______	______		______
3	______	+ oxygen	→	water
	______	______		______

Tables

Putting information into tables is a way of helping you to organise information from a textbook into your own notes. Also, thinking about whether the information is relevant and where to place it in the table helps you to remember the information and learn it!

In the following tables, you need to look carefully at the headings and then look for information that is under each heading in the textbook. If the information does not relate directly to the heading, do not include it.

The next two exercises have tables for you to organise and think about.

Exercise 5 Completing a table

After reading Wanted and unwanted reactions on page 37

In this exercise, there are simply two lists which do not relate to each other.

Wanted chemical reactions	Unwanted chemical reactions
1 ______	1 ______
2 ______	2 ______
3 ______	3 ______
4 ______	4 ______

Exercise 6 Completing and interpreting a table

After reading Oxygen and Hydrogen on pages 39–40 and Carbon dioxide on page 43

In the table below, there are three columns. There are also five questions for you to answer about each column. Look very carefully at the table before reading the textbook so that you know what you are looking for. If the information in the textbook does not answer the question about the particular gas, it is not relevant.

Questions	Hydrogen	Oxygen	Carbon dioxide
What is the chemical symbol for this gas?			
What does this gas look like?			
What are some properties of this gas?			
What percentage of Earth's atmosphere is this gas?			
How is this gas used in everyday life?			

After completing the table above:

1 Compare it with other students' tables. If there are differences, check the textbook again.

2 Look for similarities between hydrogen, oxygen, and carbon dioxide. Discuss your observations with other students.

Exercise 7 Comparing and contrasting

Comparing

Carbon dioxide is a colourless gas. Hydrogen is a colourless gas.
We can use comparison linking words to link these sentences.

both	same	in both	like
in both cases	similarly	and	as well as
in the same way	just as... so...	alike	similar in that

Both carbon dioxide and hydrogen are colourless gases.

Contrasting

Hydrogen is much lighter than air. Carbon dioxide is much more dense than air.
We can use contrast linking words to link these sentences.

on the other hand	in contrast to	different from	however
alternatively	but	although	rather than
whereas	while	unlike	yet
even though	less than	more than	whilst

Hydrogen is much lighter than air **whereas** carbon dioxide is much more dense than air.

Comparing and contrasting

We can now join our two sentences into one using *comparison and contrast linking words.*

Both carbon dioxide and hydrogen are colourless gases, however hydrogen is much lighter than air whereas carbon dioxide is much more dense than air.

1 Join the following sentences using a comparison linking word:
In a physical change, the properties of the substance change.
In a chemical change, the new substances formed have different properties.

____

____

____

2 Join the following sentences using a contrast linking word:
In a physical reaction, no new substances are formed.
In a chemical reaction new substances are formed.

____

____

____

Roundup

Exercise 8 Understanding on three levels

Level 1: Reading for accuracy
(Be able to show exactly where statements in the textbook support your answer.)

1 Is water changing to ice a physical change or a chemical reaction?

2 Which gas is heavier—hydrogen or carbon dioxide?

3 What is the difference between a reactant and a product?

Level 2: Drawing conclusions
(Be able to use information in the textbook to show how you arrived at your conclusion.)

1 If a fire extinguisher was filled with hydrogen gas, what would happen when the trigger was pressed to release the hydrogen over a fire?

2 If a fire extinguisher was filled with oxygen gas, what would happen when the trigger was pressed to release the oxygen over a fire?

3 You want to store dry ice in a freezer. What temperature would the freezer need to be?

Level 3: Applying your knowledge
(Be able to apply your knowledge in a wider context.)

1 If you put yeast in a warm sugar solution in a test tube and add a stopper, after a while the stopper will pop out. Why?

2 What would be the effect on carbon dioxide levels in the air if all the trees in a city were cut down to build more factories and highways?

3 Why do bubbles appear in soft drink only when you shake the bottle or take the lid off?

Chapter 3

Energy in our lives

Overview

ENERGY

What is it?

Energy is the ability to do work.

How is it measured?

joules (J)
kilojoules (kJ)
megajoules (MJ)

What types are there?

kinetic (movement)
gravitational potential (gravity)
elastic potential (springs)
chemical potential (chemicals)
sound (sound waves)
nuclear (atoms)
heat (hot objects)
light (light waves)
electrical

Can it be transferred?

everyday examples e.g.
rubbing hands—kinetic energy → heat energy
electric drill—electrical energy → kinetic energy, sound energy, heat energy

Where does it come from?

Renewable

Solar energy
hydro-electric energy
wind energy

Non-renewable

fossil fuels
(coal, oil, natural gas)

What do you know already?

Exercise 1 Understanding your preferred learning style

There are several different ways in which learners prefer to learn. Four types are listed below. Note that each style has advantages and disadvantages. There is no 'correct' learning style. Your task is to tick statements (both advantages and disadvantages) that apply to the way you like to learn. The learning style with the most ticks (advantages and disadvantages) is your preferred learning style.

Enthusiastic			
Advantages	✓	**Disadvantages**	✓
• I ask lots of questions. • I talk things over with others. • I am willing to take risks. • I express my feelings openly. • I use all my senses when learning.		• I don't stop to think through problems. • I am easily distracted. • I try to solve too many problems at once. • I can't be bothered doing thorough research. • I rush into quick solutions.	

Logical			
Advantages	✓	**Disadvantages**	✓
• I reason problems out step by step. • I pay attention to detail. • I work well on my own. • I am good at making plans, lists and timetables. • I stick with one problem at a time.		• I don't trust my own feelings enough. • I collect too much information before I begin a set task. • I like to follow a set way. • I don't like taking risks or trying 'silly' ideas. • I don't use other people enough.	

Imaginative			
Advantages	✓	**Disadvantages**	✓
• I see lots of alternatives. • I can daydream and visualise ideas. • I can see a pattern from many different angles. • I am good at seeing the whole picture. • I listen to others and share ideas.		• I spend too much time on working out alternatives. • I enjoy working with a problem rather than solving it. • I forget important details. • I can get distracted by other ideas. • I am not assertive enough in getting help.	

Practical			
Advantages	✓	**Disadvantages**	✓
• I get on with the task. • I want to get results. • I am good at finding the necessary information. • I don't get distracted. • I pay attention to detail.		• I don't use other people or other ideas enough. • I am suspicious about feelings—my own and others. • I don't use my imagination enough. • I think there is only one way to solve a problem—my way. • I am poor at coming up with new ideas.	

By knowing and understanding your own learning style, you can build on your strengths and consciously adopt the advantages of other styles. This will make you a much better learner.

Also, by knowing and understanding others' learning styles you can work more effectively in a group.

Exercise 2 Discussing

Form a group of three containing three different types of learners if possible. Discuss and write responses to the following questions. You should take turns at leading the discussion and writing down the group's answers.

1 If an athlete says he 'ran out of energy' during a race, describe exactly what happened in his body.

____

____

Feeling tired and listless?
Need more energy?
Buy CHOCO BOOSTER BARS
and live life to the MAX!!

CHOCO

2 Is this true or false advertising?

____

____

____

Explain your answer.

____

____

____

3 List the energy sources which produce the following results. The first one is done for you.

Energy source	Result
Petrol	a car moves
	a torch shines
	a tree grows
	a bicycle moves
	a yacht sails
	a surfboard comes down a wave
	a computer operates
	a golf ball flies in the air
	a bungee jumper falls towards the earth
	an atomic bomb explodes

Now share your ideas with other groups in your class.

During reading

Exercise 3 Clarifying ideas and concepts

After reading What is energy? on page 54

Complete the cartoon below about work and energy.

Exercise 4 Completing sentences

After reading Measuring energy on page 54 and the table on page 56

Complete these sentences.

1 The unit for measuring energy is one ______________ .

2 You use one joule of energy to lift a ______________ gram mass to a height of __________ metre.

3 One thousand joules is known as a ______________ and has the symbol __________ .

4 If you ate 20 chocolate biscuits how many km would you have to walk to 'walk it off'? __________

Exercise 5 Recalling information

After reading Forms of energy on pages 57–59

Some words have a slightly different meaning in science to what they have in everyday use. For example, the everyday meaning of *work* is physical or mental effort directed towards doing or making something. The scientific meaning is slightly different. *Work* is the result of a force moving an object a certain distance.

Scientific words often have a describing word in front of them; for example, 'kinetic energy'. Kinetic comes from the Greek word *kinetikos*, which means 'to move'. So kinetic energy is the energy that a moving object has. There may even be two describing words, as in the first exercise below.

1 Fill in the missing words below to give the meaning of gravitational potential energy.

a Potential means stored. Gravitational means to do with gravity.

So gravitational potential energy is ______________ energy that a body has because of ______________________

____

b Elastic means stretched. Potential means stored. So elastic potential energy is ______________________

2 Philippa finds the graphic on the right useful for reminding her what gravitational potential energy is. Explain what you think the graphic means.

O
P TENTIAL

____

____

____

____

3 Fill in the three gaps in the teacher's whiteboard summary.

ENERGY CLASSIFICATION KEY
ENERGY
POTENTIAL
....
GRAVITATIONAL
....
A stretched rubber band has a particular type of energy, called ....

4 a Write a definition for kinetic energy.

____

b Why do you think sound is considered to be a type of kinetic energy?

____

____

____

5 Complete the table below by listing how each form of energy is stored.

Form of energy	Stored in
chemical energy nuclear energy elastic potential energy gravitational potential energy	

6 Linda is enjoying a swing at the local park. Complete the following sentences describing her motion.

a Linda's speed is greatest at position __________

b Linda has most kinetic energy at position __________

c Linda has most gravitational potential energy at position __________

1
2
3

Exercise 6 Labelling

After reading Energy changes on page 60

After visiting a coal-burning power station, Clayton drew this diagram of how it works. Add the following labels to the diagram.

electrical energy	kinetic energy	stored chemical energy

Note that you will have to use one label twice.

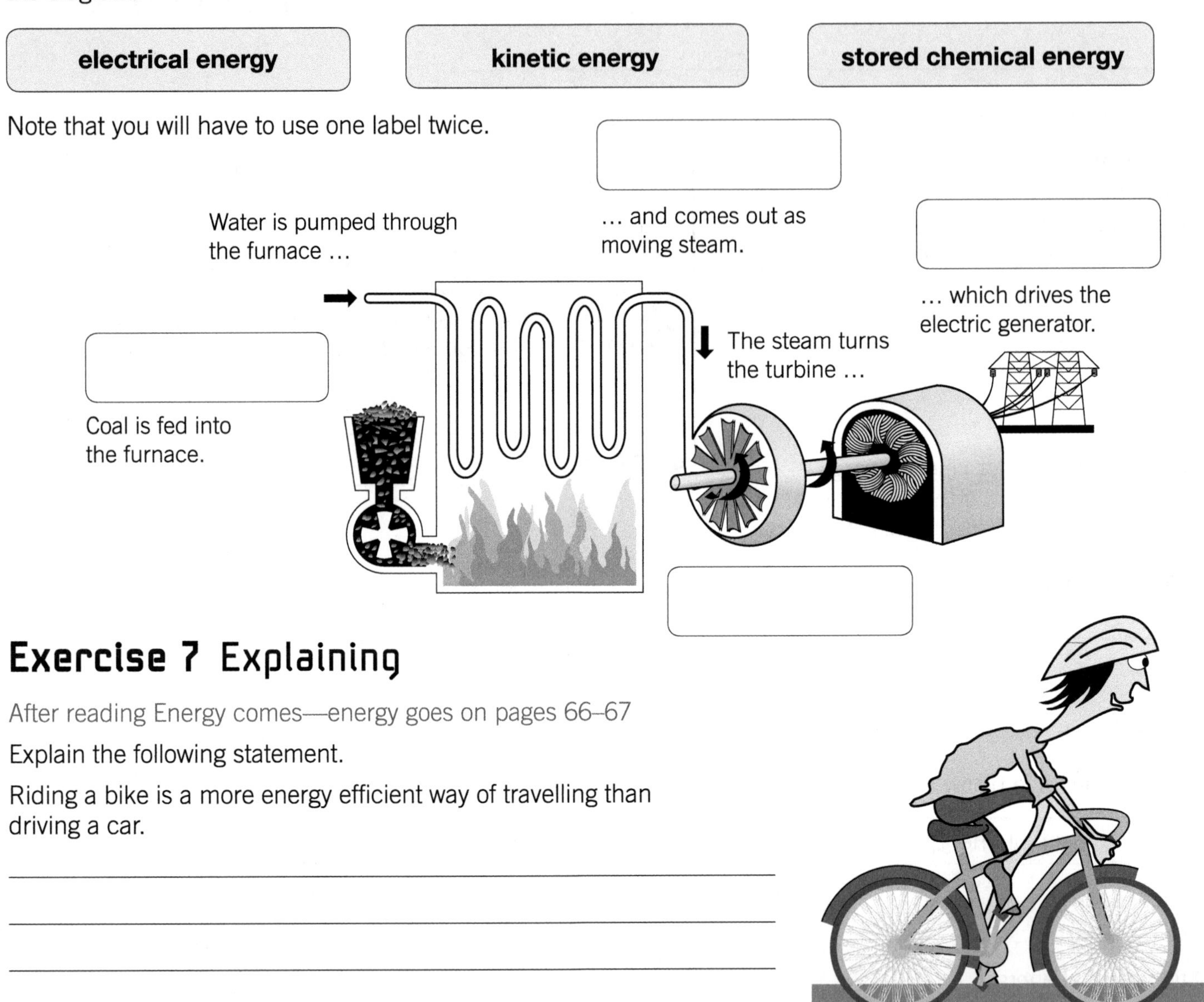

Exercise 7 Explaining

After reading Energy comes—energy goes on pages 66–67

Explain the following statement.

Riding a bike is a more energy efficient way of travelling than driving a car.

____

____

____

____

Exercise 8 Creating an analogy

After reading Conservation of energy on page 67

An analogy is a comparison used to show similarity.

In the textbook, the authors have used Monopoly as an analogy to help you understand the concept that the total amount is the same even though it is distributed differently.

Create another analogy using an everyday example which explains the law of conservation of energy.

____

____

____

Exercise 9 Understanding on three levels

After reading Where does energy come from? on pages 68–69

For each of the statements, write T (true) or F (false) in the brackets provided. When you have finished, discuss your answers in small groups. Try to reach an agreement on your answers. You must be ready to give reasons for your answers.

Level 1: Reading for accuracy

(Be able to show where statements in the textbook support your answer.)

1 Animals get their energy by eating plants (or eating other animals that have eaten plants). (   )

2 Photosynthesis is the process in which green plants are eaten by animals. (   )

3 Fossil fuels are made from the buried remains of ancient plants and animals. (   )

4 Hydro-electricity is a non-renewable energy source. (   )

Level 2: Drawing conclusions

(Be able to use information in the textbook to show how you arrived at this conclusion.)

1 The sun is responsible for most of the forms of energy on the Earth. (   )

2 It takes millions of years for fossil fuels to be made. (   )

3 Both coal and oil were formed as a result of great pressure. (   )

4 Solar energy is converted into chemical energy in an animal's body. (   )

Level 3: Applying your knowledge

(Be able to apply your knowledge in a wider context.)

1 Eventually fossil fuels will run out. (   )

2 Power stations are 100 per cent efficient. (   )

3 Hydro-electricity is the best form of renewable energy. (   )

4 If the sun stopped shining, life could not exist on Earth. (   )

Roundup

Exercise 10 Applying ideas and concepts

See if you can apply what you have learnt in this chapter by explaining how one (or more) of the following devices works. Make sure you talk about forms of energy and energy changes. (You may find some of the *cause and effect linking words* on page 92 of this workbook are useful in describing how the device works.)

1 When the weight falls the bulb glows.

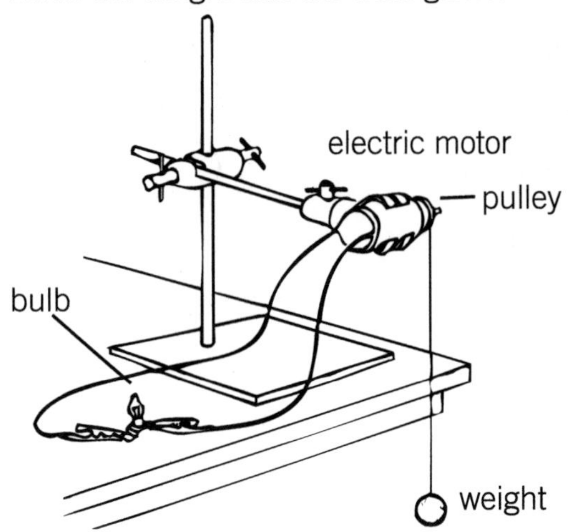

2 The steam turns the pin-wheel.

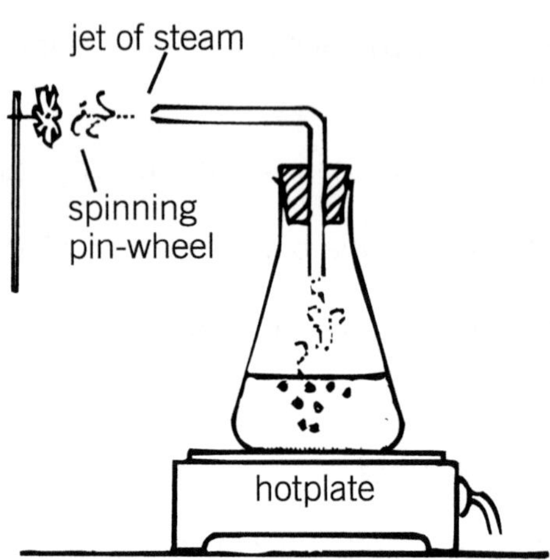

3 The falling water turns the wheel.

4 Spin the wheel and the bell rings.

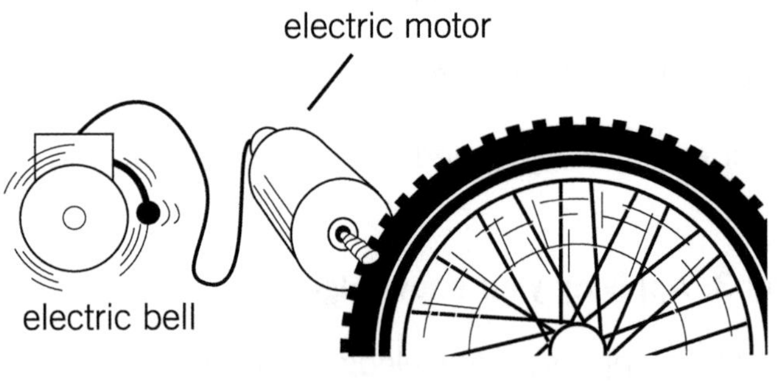

My explanation of device number ___________

____

____

____

____

____

____

____

____

Chapter 4

Growth and reproduction

Overview

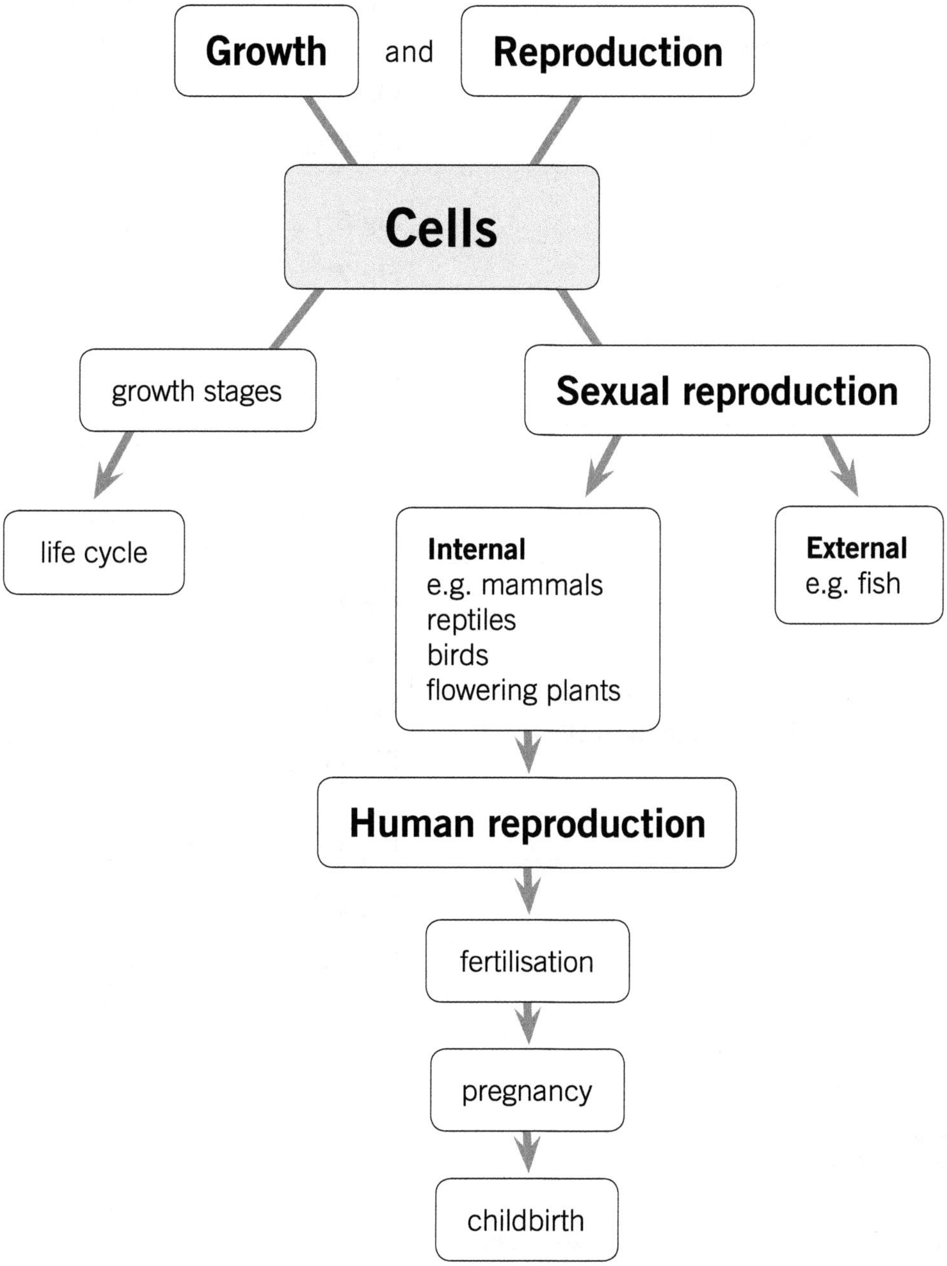

Growth and **Reproduction**

Cells

growth stages

life cycle

Sexual reproduction

Internal
e.g. mammals
reptiles
birds
flowering plants

External
e.g. fish

Human reproduction

fertilisation

pregnancy

childbirth

What do you know already?

Exercise 1 Completing a graphic outline

Each chapter in your textbook has been organised by the authors into headings and sub-headings, figures and investigations. In this exercise, you will be exploring the chapter to find out how the written and visual information has been organised. The first section of the chapter is done for you as an example.

Main Headings	Sub-headings	Figures
4.1 Growth	**How bones grow** **Growth stages** **Life cycles**	**Fig 1** Cells divide to make new cells **Fig 2** X-ray of one-year-old child **Fig 3** X-ray of 12-year-old child **Fig 4** Life cycle of a frog **Fig 5** Life cycle of a beetle
4.2 Producing new life	________________ ________________ ________________ **The menstrual cycle**	**Fig 6** ________________ **Fig 7** ________________ **Fig 8** ________________ **Fig 9** ________________ **Fig 10** ________________ **Fig 11** ________________ **Fig 12** ________________ **Fig 13** ________________ **Fig 14** ________________ **Fig 15** ________________ **Fig 16** ________________
4.3 ________________ ________________	________________ ________________ ________________	**Fig 17** ________________ **Fig 18** ________________ **Fig 19** ________________

During reading

In this chapter, you will find Core and Challenge exercises.

After reading each section of the textbook, you will find an exercise to complete. If your solution to this exercise is correct (check your answers on page 32 of this workbook), go straight to the Challenge exercise. If you cannot do the exercise, or you get it wrong, or even if you get it right but are not sure about it, read the textbook section again and go to the Core exercise. Of course, you may then complete the Challenge exercise.

Exercise 2

After reading Growth on pages 78–82

Recalling information – multiple choice

Circle the correct answer.

a Approximately how many cells are there in your body?

(i) 50 000 000 (ii) 50 000 000 000 (iii) 50 000 000 000 000

b The part of the cell which controls the activities of the cell.

(i) Cytoplasm (ii) Nucleus (iii) Brain

c Old age is regarded as being

(i) over 20. (ii) over 50. (iii) over 75.

d A frog's life cycle has

(i) 5 stages. (ii) 3 stages. (iii) 4 stages.

e If you compare a frog's egg with a human egg, the frog's egg is

(i) smaller. (ii) larger. (iii) about the same size.

Check your answers on page 32 of this workbook.

Core exercise

Match the word with the meaning

Fertilisation	A jelly-like substance that fills most of a cell.
Adolescence	The process in which the nucleus of a sperm and ovum join to make a new living thing.
Puberty	The 'grub' stage in the life cycle of an insect.
Metamorphosis	The period in the life cycle when sexual development occurs.
Larva	A dramatic change in an organism's appearance and habits during a life cycle.
Cytoplasm	Stage of a human life between childhood and adulthood.

Challenge exercise

Growth occurs when certain cells *multiply*. Why is this process known as *cell division*?

Why are the hen's eggs much larger than human eggs?

Exercise 3

After reading Producing new life on pages 83–88

Recalling information – True or False

Circle the correct answer and be prepared to explain your response.

1 Large organisms reproduce asexually and small simple organisms reproduce sexually. TRUE FALSE

2 Ova are made in ovaries and sperm cells are made in testes. TRUE FALSE

3 Males generally start puberty before females. TRUE FALSE

4 Females have ova when they are born but males do not have sperm until they reach puberty. TRUE FALSE

5 A foetus is smaller than an embryo. TRUE FALSE

6 Human females are the only mammals to have a menstrual cycle. TRUE FALSE

Check your answers on page 32 of this workbook

Core exercise

Place the following words in the table where relevant. **Note:** some words will apply to both males and females.

ovary, sperm, testes
menstruation, puberty, vagina
penis, prostate, uterus
semen, placenta, ovulation
ejaculation, pubic hair

Human male	Human female

Challenge exercise

What causes you to have a belly button?

Investigate the effect the rubella virus can have on a developing baby.

Exercise 4

After reading pages 90–91

Making decisions from alternatives

Cross out the incorrect word in each sentence.

1 When eggs are fertilised outside the female's body, it is known as (internal/external) fertilisation.

2 If external fertilisation occurs, the young are more likely to be (dependent on/independent of) their parents.

3 The fewer the eggs produced, the more likely that the young will be (dependent on/ independent of) their parents.

4 Even though the eggs of sea turtles are fertilised (internally/externally) the young are (dependent on/independent of) their parents when hatched.

5 The eggs of a seahorse are produced by the (male/female) and the fertilised eggs are carried until hatched by the (male/female) in a pouch.

Check your answers on page 32 of this workbook.

Core exercise

Compiling tables

The following animals are mentioned in your textbook. Place them where they belong in the tables below.

frogs	birds	wombats
turtles	fish	seahorses

Internal reproduction	External reproduction

Dependent when hatched/born	Independent when hatched/born

Challenge exercise

Completing a table

Examine the table at the top of page 91 of your textbook. Use your knowledge of human reproduction and behaviour to complete the following column.

	Human (mammal)
Number of eggs produced each year	
How eggs are fertilised	
Parental care	

Exercise 5

After reading Reproduction and survival in flowering plants on pages 91–94

Matching words and meanings

Write each word in the shaded box next to its meaning below.

dispersal	vegetative reproduction	stigma
pollen	sexual reproduction	anther
sepal	ovary	fruit

1 The sticky female part of a flower on which pollen lands. __________

2 The male organ in a flower which makes pollen. __________

3 Movement of a plant's seeds away from the plant. __________

4 Part of the flower which contains the egg cell. __________

5 The process in which plants reproduce from parts of the adult plant. __________

6 Mature, fertilised ovules of a plant. __________

7 Fine, powdery substance containing sperm. __________

8 The firm, outer case of a flower. __________

9 The process in which male sperm and female eggs produce offspring. __________

Check your answers on page 32 of this workbook.

Core exercise

Sequencing

Use the numbers 1–6 to place the following steps in the correct sequence.

☐ Petals, sepals and stamen wither and fall off.

☐ Pollen then fertilises the ova in the ovary.

☐ Pollen containing sperm lands on the stigma.

☐ The ovary becomes a fruit.

☐ Seeds develop in the ovary.

☐ Pollen tubes grow down to meet the ovary.

Challenge exercise

Making decisions from alternatives

Tick the statement in each pair which is correct.

1 Flowering plants reproduce sexually.
Flowering plants reproduce asexually.

2 Stigma and ovaries are female organs.
Stigma and ovaries are male organs.

3 Pollen travels down a tube which joins the stigma to the ovary.
Pollen grows its own tube to carry the sperm to the ovary.

4 A tomato is a fruit.
A tomato is a vegetable.

5 All fruits contain seeds.
Not all fruits contain seeds.

Roundup

Exercise 6 Writing and speaking in role

Your task
- Complete the script below and then record it on a tape recorder or perform it for your class.

Your roles
- Work in pairs.
- One of you is a famous biologist whose area of expertise is plant and animal reproduction. Think of a name for yourself.
- The other person is a radio host who has absolutely no idea about plant and animal reproduction. Think of a name for yourself.

Host: Good evening listeners. Welcome to station __________ . This is your favourite radio interviewer, ________________, and I am very pleased to have with me here tonight the world famous biologist, ________________ . He/she will be telling us everything we ever wanted to know about cells and reproduction. Welcome, ________________ .

Biologist: Thank you ________________, it's nice to be here.

Host: Well, let's get right into it. Plants and animals reproduce in the same way, don't they?

Biologist: ____

____

____

____

Host: ____

Biologist: ____

____

____

____

Host: ____

Biologist: ____

____

____

____

Host: Wow! Thank you ________________, I had no idea it was so interesting and so complicated. Isn't nature wonderful?

Answers

Exercise 2

a (iii) 50 000 000 000 000

b (ii) Nucleus

c (iii) over 75.

d (ii) 3 stages.

e (ii) larger.

Exercise 3

1 FALSE

2 TRUE

3 FALSE

4 TRUE

5 FALSE

6 FALSE

Exercise 4

1 external

2 independent of

3 dependent on

4 internally/independent on

5 female/male

Exercise 5

1 stigma

2 anther

3 dispersal

4 ovary

5 vegetative reproduction

6 fruit

7 pollen

8 sepal

9 sexual reproduction

Chapter 5
Particles

Overview

THE PARTICLE MODEL

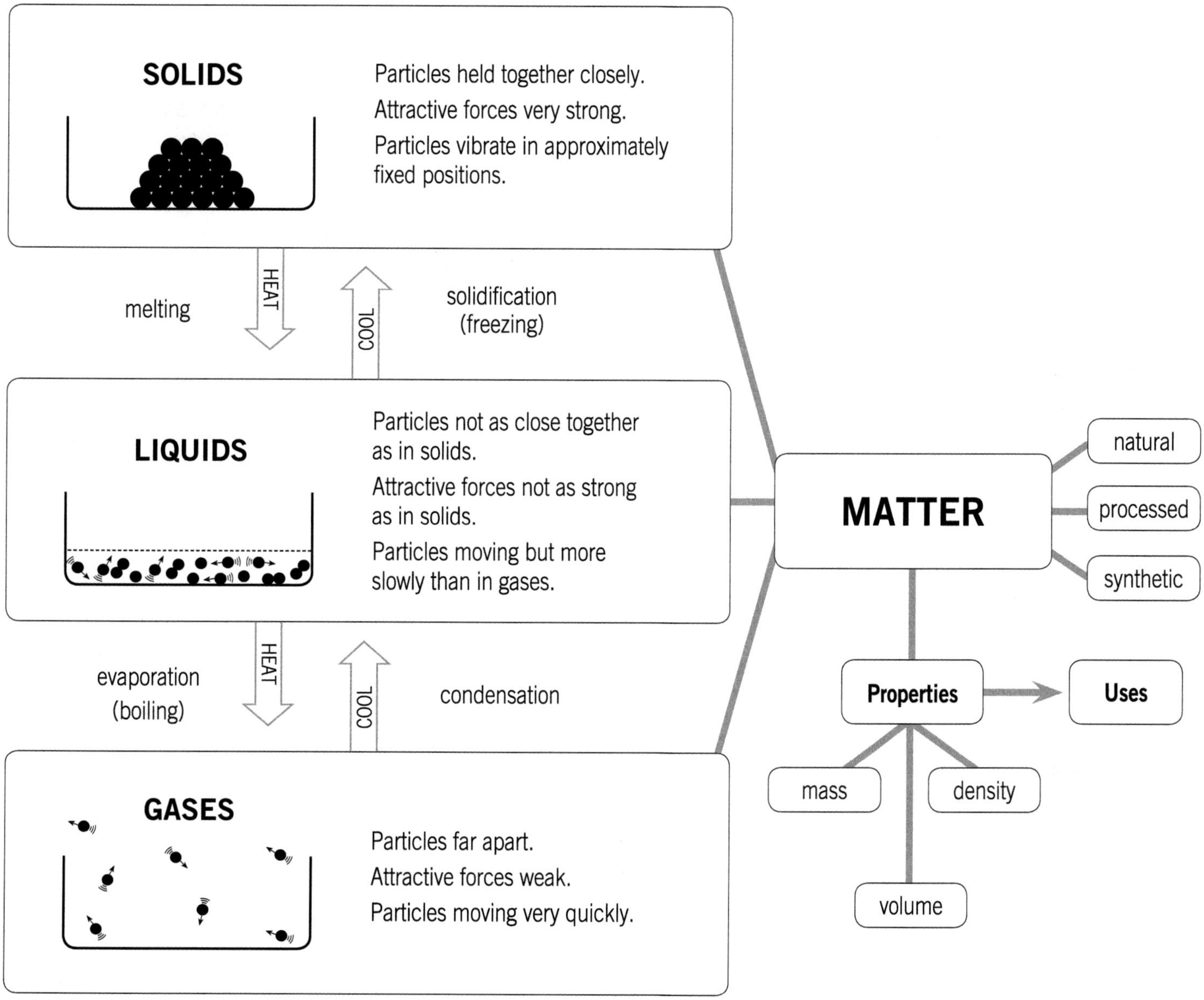

SOLIDS

Particles held together closely.

Attractive forces very strong.

Particles vibrate in approximately fixed positions.

melting — HEAT

COOL — solidification (freezing)

LIQUIDS

Particles not as close together as in solids.

Attractive forces not as strong as in solids.

Particles moving but more slowly than in gases.

evaporation (boiling) — HEAT

COOL — condensation

GASES

Particles far apart.

Attractive forces weak.

Particles moving very quickly.

MATTER

- natural
- processed
- synthetic

Properties → **Uses**

- mass
- density
- volume

What do you know already?

Exercise 1 Understanding words in context

In this chapter, you will need to understand the meanings of the words in *italics* in the sentences in the first column of the table below.

Before you look up the meaning in the **Glossary** section at the back of your textbook or in a dictionary, write in your own words what you think the meaning of the words might be. Use the clues from the other words around them to try to work out a meaning that makes sense. The first one is done for you as an example.

Words in context	Possible meaning	Glossary or dictionary meaning
Everything around you is made up of *matter*.	Stuff	Anything that has mass and occupies space.
The *density* of gold is greater than the *density* of feathers.		
When water vapour *condenses* it forms clouds.		
When you open a bottle of perfume the fragrance soon *diffuses* throughout the room.		
Cotton is a natural product but nylon is *synthetic*.		

During reading

Exercise 2 Summarising

After reading Properties of matter on page 100

In the Activity on page 101 you used a table to summarise the properties of solids, liquids and gases.

1 Complete this sentence comparing and contrasting solids and liquids.

Solids and liquids have similar ____________, but liquids do not have a ____________ ____________ .

2 Now it's your turn. Write a sentence comparing and contrasting liquids and gases. Use *comparison and contrast linking words* from page 92 of this workbook.

____

____

Exercise 3 Clarifying ideas and concepts

After reading Density on pages 102–103

Complete the cartoons below.

Exercise 4 Creating a table

After reading Using materials on page 104 and Synthetic materials on page 105

1 Use all the space below to create a table with three columns and three rows.

2 Insert the following headings for each column in the first row.

natural materials processed materials synthetic materials

3 Select the correct description for each heading and place it in the second row under its heading.

- Materials in their natural state.
- Materials which do not occur naturally.
- Materials which occur naturally but whose properties have been improved or altered.

4 In the third row of your table place the materials listed below into the relevant column.

- leather
- glass
- WaterSorb
- paints
- rubber
- pesticides
- aluminium
- cotton
- bread
- polystyrene
- wood
- Sunsorb
- paper
- silk
- seafood
- salt
- wool
- plastic
- argon
- gold nugget

Exercise 5 Completing a table

After reading Solid—liquid—gas on page 108

1 Complete the table by filling in the blank spaces.

Change of state	Heating or cooling	Action/doing word (verb) to describe change	Naming word (noun) to describe change
ice ⟶ water	heating	melt	melting
metho ⟶ metho vapour		evaporate	
water ⟶ steam			boiling
water vapour ⟶ water	cooling		
molten gold ⟶ gold bar			
			freezing

2 In one sentence, explain the difference between evaporation and boiling. Use a *contrast linking word* from page 92 of this workbook.

____

____

Exercise 6 Observing, inferring and predicting

After reading Solid—liquid—gas on page 108

As Tracey watches the electric kettle boil she has five different thoughts (below). Label these thoughts as either observation, inference or prediction. (You may need to check Chapter 2 of this workbook to remind yourself of the differences between these thinking skills.)

Exercise 7 Contrasting

After reading The particle theory on pages 109–112

1 Complete the table below which summarises the differences between solids, liquids and gases in terms of the particle theory.

	Distance between particles	Forces between particles	Speed of particles
SOLIDS			
LIQUIDS			
GASES			

2 Use the information in the table above to write a paragraph contrasting (describing differences between) solids, liquids and gases. Use *contrast linking words* from page 92 of this workbook.

STEP 1 Write a general statement (a topic sentence) about solids, liquids and gases.

STEP 2 Write a sentence or two contrasting the distances between particles.

STEP 3 Write a sentence or two contrasting the forces between particles.

STEP 4 Write a sentence or two contrasting the speed of the particles.

Exercise 8 Interpreting illustrations

After reading Diffusion on pages 117–118

The illustrations in a textbook are just as important as the text itself—sometimes more important. They may be diagrams, cartoons or photos.

Look at the three-part diagram on page 117 showing how perfume diffuses. Find three sentences in the text that are relevant to the diagram. Write these sentences into the boxes below to describe each part of the diagram.

Exercise 9 Inferring and illustrating

After doing the Activity on page 118 and reading page 119

1 Look at Part C of the Activity on page 118.

  a Sketch the second part of the illustration below to show what happened.

  b Write a caption that describes what your two-part illustration shows.

Your sketch goes here

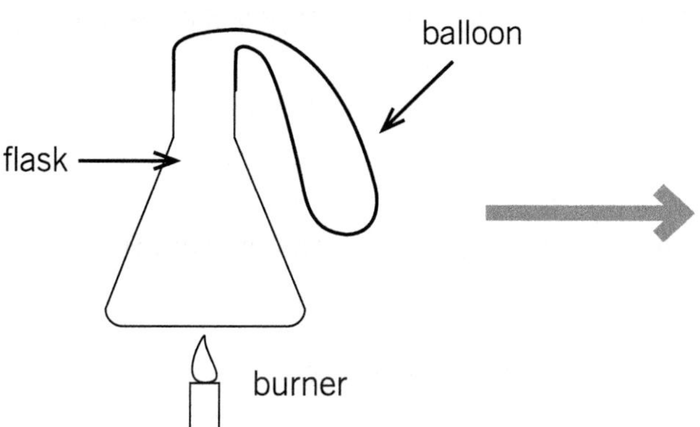

Caption: ____________________

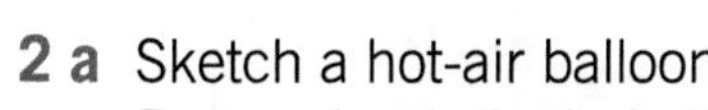

2 a Sketch a hot-air balloon travelling through the air. Remember to include the basket and burner.

  b Sketch the air particles inside and outside the balloon.

3 Infer what happens to the air particles inside the balloon when the burner is turned on.

____

____

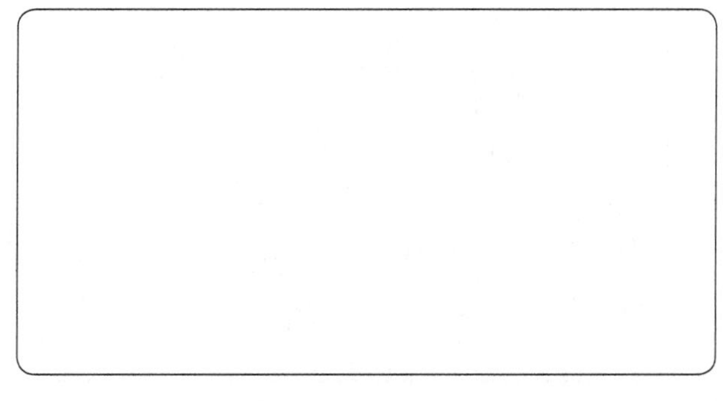

Roundup

Exercise 10 Clarifying ideas and concepts

Complete the cartoon captions.

Exercise 11 Conducting a class quiz

1 Select a quiz master and a scorer then divide the rest of the class into two teams (Team A and Team B).

2 Distribute a blank card to each student. On the card, each student is to write a Chapter 3 question (and an answer on the other side).

3 Team A students place their cards in the Team A box.

4 Team B students place their cards in the Team B box.

5 The quiz master selects a question card from the Team A box and asks the question of Team B.

6 If Team B members get the answer correct, the scorer awards one point to Team B.

7 Team B keeps answering questions until they get one incorrect. When they are incorrect, it is now Team A's turn to answer questions.

8 When all question cards have been selected, the team with the highest score is the winner.

Chapter 6

Cells

Overview

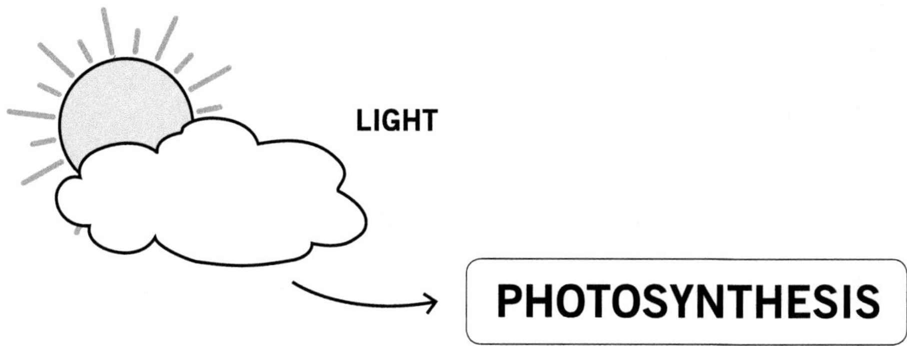

PHOTOSYNTHESIS

carbon dioxide + water + energy ⟶ glucose + oxygen

CELL

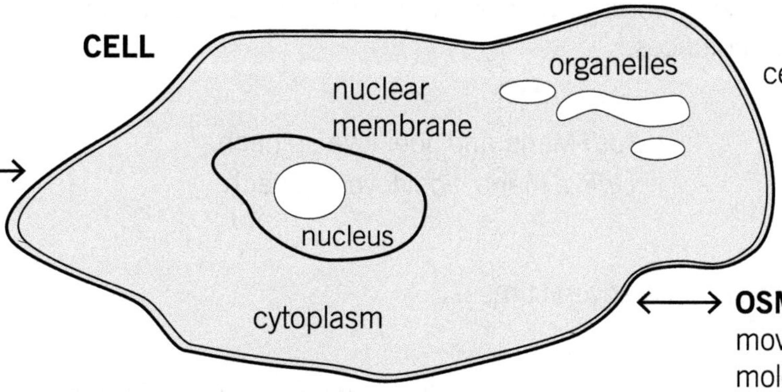

cell wall

organelles

nuclear membrane

nucleus

cytoplasm

DIFFUSION ⟷
movement of molecules in to and out of cells

⟷ **OSMOSIS**
movement of water molecules through cell membranes

CELL RESPIRATION

glucose + oxygen ⟶ carbon dioxide + water + ENERGY

ENERGY ⟶ cell division; muscle movement

cell division ⟶

CONTROLLED
- growth
- replacement

UNCONTROLLED
- cancer

What do you know already?

Exercise 1 Reflecting and sharing

Think about each of these questions individually before joining a small group to share your ideas with others.

1 Which is smallest—a cell, an atom, or a molecule?

2 What is the most common type of cancer in Australia?

3 Which type of cancer causes the most deaths in Australia?

4 What causes your muscles to become tired after vigorous exercise?

5 Why are anabolic steroids banned in all sports?

6 Why does a cube of potato become soft in salty water but hard in distilled water?

Keep your answers in mind as you work through the chapter. You will review your answers in exercise 7.

Exercise 2 Understanding linking words

Understanding the meaning of linking words is vital when reading. Look at the difference the linking words make to the meaning of the following sentences.

> *Both* Maria and Joel love spinach.
> *Unlike* Maria, Joel loves spinach.

Read this paragraph about cells carefully several times.

> Cell division is usually an orderly process which occurs in growth regions of the body, such as bones, or in regions, such as the skin, that specialise in cell replacement. However, in cancer this orderly process goes wrong. Cancer occurs when certain cells divide rapidly and uncontrollably. The cancer cells look different from the normal cells in the area and they do not function like normal cells. For example, in liver cancer, cancer cells grow rapidly and replace normal liver cells. Gradually the liver loses its function and death usually occurs.

1 Find a generalisation linking word. ______________________________

2 Find two sequence linking words. ______________________________

3 Find two contrast linking words. ______________________________

4 Find two types of linking words which show that examples will follow.

______________________________ ______________________________

Exercise 3 Defining

In Science, you will be asked to define words. This language skill has its own structure and is different from describing or explaining. It is easy to define when you know the structure for expressing your knowledge. Here are some definitions. Find a common pattern for writing a definition.

A geologist is a scientist who studies the Earth's crust.

A bird is an animal that has constant body temperature and has feathers. An example is a chicken.

A centrifuge is a machine which separates mixtures by a spinning motion. An example is a spin dryer.

A definition is a group of words which has the following pattern:

A **(word to be defined)** is a **(larger group to which it belongs)** that/which/who **(list the characteristics which make the word different from the rest of the group)**.

An example is ______________________ (*an example may not need to be added*).

Lets define the word **volcano**.

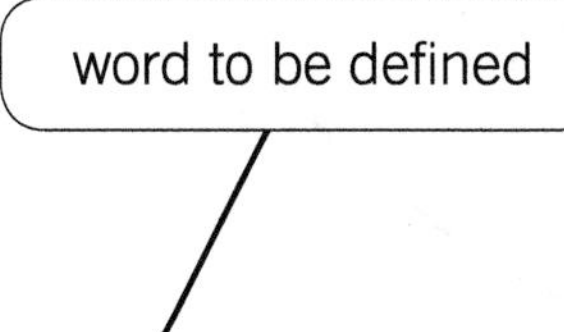

word to be defined

larger group to which it belongs

characteristics which make volcanoes different from other mountains

an example

A **volcano** is a **mountain** which **is formed when pressure inside the Earth pushes magma through weak spots on the Earth's crust**. An example is **Mt Vesuvius**.

Using this pattern write a definition for:

a A dinosaur ______________________

____

____

b Muscles ______________________

____

____

During reading

Exercise 4 Recalling information

Follow the directions on the path and answer the questions in your notebook or on a separate piece of paper. Then turn to page 46 of this workbook to correct your answers.

Do not go to exercise 5 unless all your answers are correct or you know why you were incorrect.

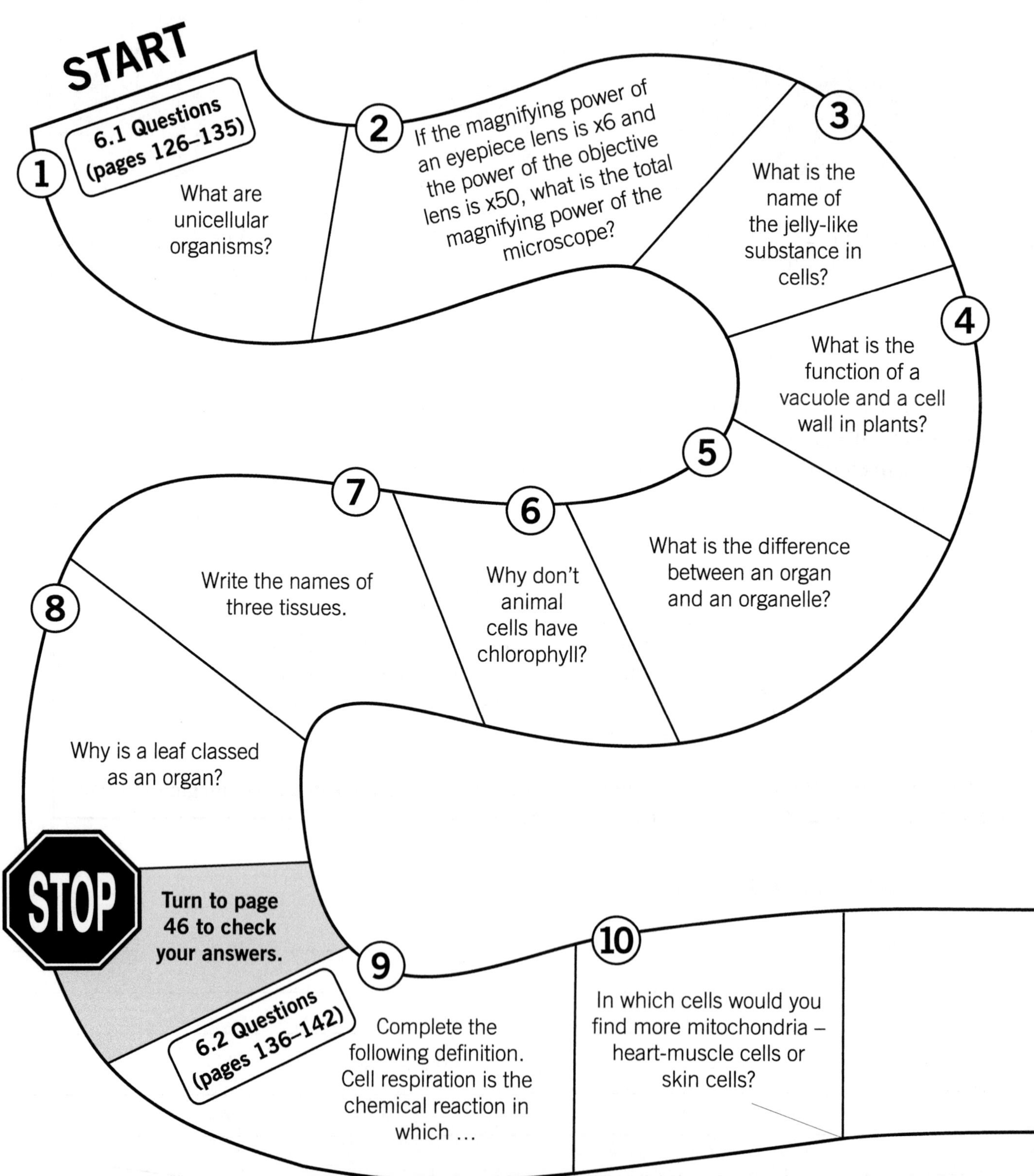

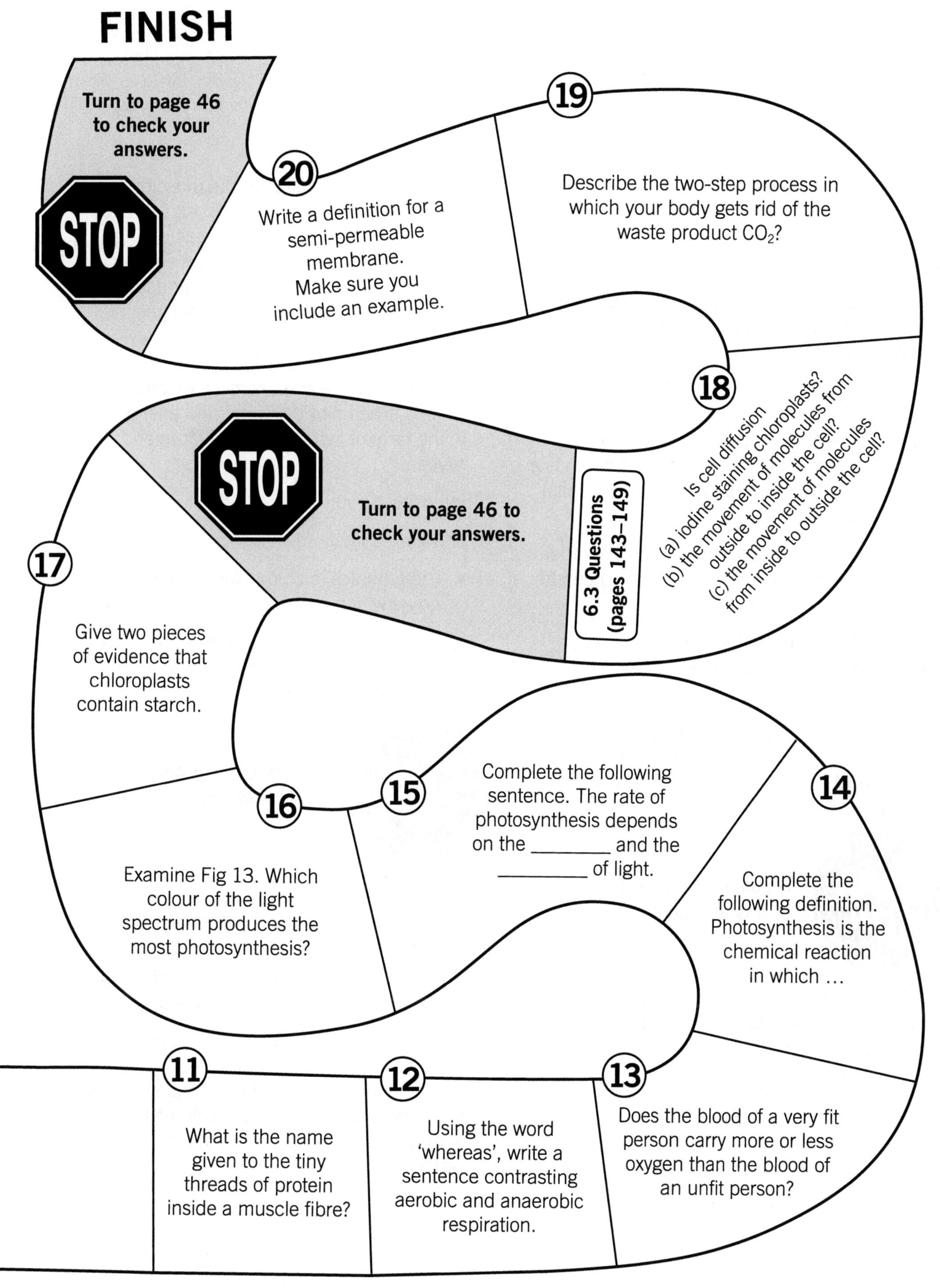

FINISH
Turn to page 46 to check your answers.
STOP
20
Write a definition for a semi-permeable membrane. Make sure you include an example.
19
Describe the two-step process in which your body gets rid of the waste product CO_2?
18
Is cell diffusion
(a) iodine staining chloroplasts?
(b) the movement of molecules from outside to inside the cell?
(c) the movement of molecules from inside to outside the cell?
6.3 Questions (pages 143–149)
STOP
Turn to page 46 to check your answers.
17
Give two pieces of evidence that chloroplasts contain starch.
16
Examine Fig 13. Which colour of the light spectrum produces the most photosynthesis?
15
Complete the following sentence. The rate of photosynthesis depends on the ________ and the ________ of light.
14
Complete the following definition. Photosynthesis is the chemical reaction in which …
11
What is the name given to the tiny threads of protein inside a muscle fibre?
12
Using the word 'whereas', write a sentence contrasting aerobic and anaerobic respiration.
13
Does the blood of a very fit person carry more or less oxygen than the blood of an unfit person?

Answers

1 Single cells that are complete.

2 x 300

3 cytoplasm

4 A vacuole is the part of the cell where water and dissolved substances are stored. A cell wall is a thick, tough layer that protects the inside of the cell and also provides stiffness for support.

5 An organelle is a structure inside a cell that helps to keep it working, whereas an organ is a collection of tissues (made up of similar cells) that has a very specific function.

6 All animals, even microscopic ones, eat other animals or plants for food, and do not need to make their own by photosynthesis.

7 Muscle tissue, nerve tissue, transport tissue, food making tissue, lining tissue, support tissue.

8 An organ is a collection of specialised tissues that have a particular function and a leaf's specialised function is to make food so it is an organ.

Go back to page 44 of this workbook.

9 Cell respiration is the chemical reaction in which glucose and oxygen combine to produce energy, carbon dioxide and water.

10 heart-muscle cells

11 fibrils

12 During aerobic respiration the muscles are supplied with oxygen and glucose is broken down to produce energy, whereas during anaerobic respiration there is not enough oxygen supplied to the muscles and lactic acid is produced which causes fatigue.

13 more

14 Photosynthesis is the chemical reaction in which carbon dioxide, water and energy in the form of light produce glucose and oxygen.

15 intensity; colour

16 orange-red light

17
- The granules in chloroplasts stain with iodine which indicates they contain starch.
- The granules disappear when plants are kept in the dark. (Photosynthesis needs light to produce starch.)

Go back to page 44 of this workbook.

18 both (b) and (c)

19 Step 1: The CO_2 diffuses out of the muscle cells into the blood.
Step 2: The CO_2 is transported by blood to the lungs and breathed out. (See page 145.)

20 A semi-permeable membrane is a membrane through which small molecules can pass but larger ones cannot. An example is a cell membrane which allows water to pass freely across it but slows down the movement of dissolved substances.

Roundup

Exercise 5 Viewing a problem from different perspectives

Scientists and people of all ages know that the carcinogens in cigarettes cause lung cancer. Yet people continue to smoke and young people are continuing to take up smoking. Why? How can we prevent young people from choosing to smoke?

One strategy to solve a problem is to look at the problem from many different points of view. Form groups of nine students with each student choosing one of the characters listed below.

- a young person who has taken up smoking
- a young person who has never smoked
- a person who has been smoking for more than twenty years
- an adult who has never smoked
- a person who is dying of lung cancer (or a relative of a person who is dying or has died of lung cancer)
- a surgeon who operates on patients with lung cancer
- a tobacco grower
- an advertising/marketing person
- the chairman of a large cigarette company

When you have decided upon your roles, think about the point of view of that character.
Now form a circle and improvise (create, without previous planning) dialogue that would occur if these characters were asked to discuss the topic of smoking.
After this role play, state your two main recommendations for solving the problem.

Solution 1 ____

____

____

____

____

____

____

Solution 2 ____

____

____

____

____

____

____

Exercise 6 Recalling words and meanings

In this chapter, there were many scientific words for you to learn and to understand. Complete the table below to help you to reinforce your understanding.

	Meaning	Word
1	To use a ruler to measure the field of view of a microscope	c _ _ _ b _ _ te
2	Small structures in cells which have specific functions	or _ _ _ _ _ _ _ s
3	Thin tissue that covers, lines or connects cells	m _ _ b _ _ _ _
4	Jelly-like substances which fills most of a cell	c_ _ _ p _ _ _ _
5	The thin protective outer layer of the skin	e _ _ _ _ r _ _ _
6	A plant organ whose main function is to make food	_ _ a _
7		vacuole
8		nucleus
9	An organ whose function is to break down food	s _ _ _ _ _ h
10	Tiny organelles in which cell respiration occurs	m _ _ _ c _ _ _ _ _ _ _
11	Tiny threads of protein in muscle fibres	f _ _ r _ _ s
12	A type of respiration that occurs without air	a _ _ er _ _ _ _
13		photosynthesis
14		chloroplasts
15		diffusion
16		osmosis
17		semi-permeable

Exercise 7 Reviewing

Go back to exercise 1 to check whether your answers were correct.

Chapter 7

Investigating heat energy

Overview

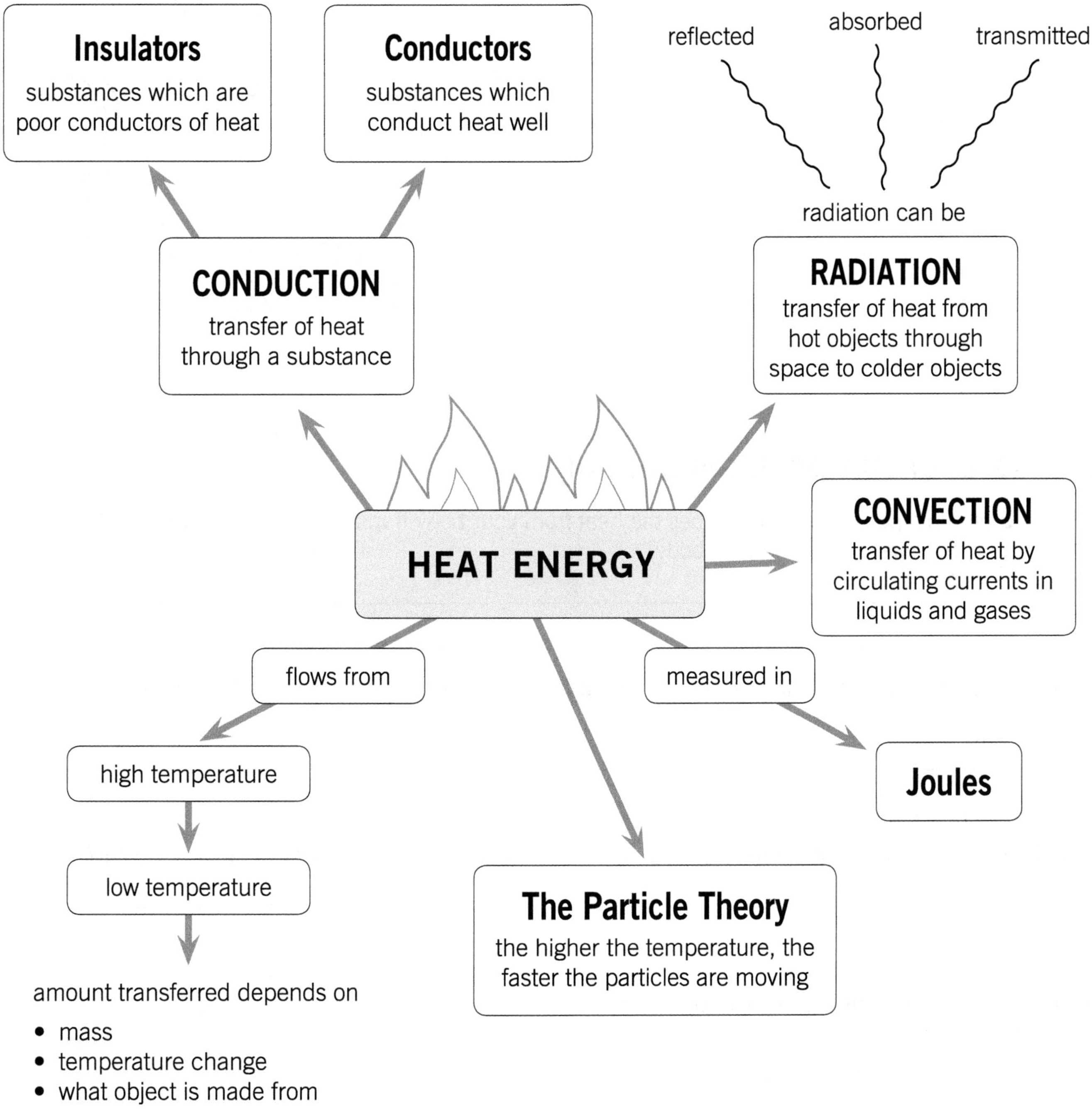

Insulators
substances which are poor conductors of heat

Conductors
substances which conduct heat well

reflected
absorbed
transmitted

radiation can be

CONDUCTION
transfer of heat through a substance

RADIATION
transfer of heat from hot objects through space to colder objects

HEAT ENERGY

CONVECTION
transfer of heat by circulating currents in liquids and gases

flows from

measured in

high temperature

Joules

low temperature

The Particle Theory
the higher the temperature, the faster the particles are moving

amount transferred depends on

- mass
- temperature change
- what object is made from

What do you know already?

Exercise 1 Observing and listing

Walk around your home and note at least 20 things that produce heat when used. (Remember, any object with moving parts will produce heat.) The moving parts do not have to be powered by electricity.

Hm! This fan is cooling – but it does have moving parts!

Exercise 2 Group problem solving

1 When you dive into a pool on a hot day, does the heat from your body transfer to the cooler pool or does the cooler temperature of the water transfer to your body?

2 Why does a plastic container melt when placed in a gas oven, but does not become hot when placed in a microwave oven?

3 On a hot day, why is the temperature in a black car much higher than the temperature in a white car?

Now share your ideas with other groups in your class.

During Reading

Exercise 3 Locating information

In this chapter of the workbook you will be asked questions which require you to locate information from throughout the whole chapter of the textbook.

Before you try answering these questions, you should:

1 Scan the whole of the chapter and make yourself familiar with how the authors have organised the information.

2 Read the headings and sub-headings. They are signposts to finding the information you need.

Complete the table below. In the first column, you are given a list of headings and sub-headings used in the textbook. Read the questions in the second column and decide which of the headings or sub-headings would be the best place to locate relevant information. Write these headings in the third column. (**Do not answer the questions**. Your task is simply to decide where to look for the answers.)
The first one is done for you as an example.

Headings and sub-headings used in the textbook Chapter 7	Questions	I would expect to find the answer in
7.1 Heat and temperature	1 List three ways in which heat can be transferred.	**7.2 Heat transfer**
Heat and the particle theory	2 How do convection currents cause sea breezes?	
7.2 Heat transfer	3 Can radiation travel through a vacuum?	
Conduction	4 Which colour absorbs the most heat?	
Convection	5 What is the difference between heat and temperature?	
Radiation	6 Which substances are good conductors?	
Absorbing and emitting radiation	7 How can you prevent substances losing or gaining heat?	
Controlling heat transfer	8 What happens to the particles in a substance as it cools down?	
7.3 Heat in everyday life	9 What do I need to consider when building a house to keep power bills as low as possible?	

Exercise 4 Understanding text

When you have located the section containing the information you need, you should read it carefully to understand the way the information is organised.

The box on the right contains a section from the textbook.

The paragraph numbers, circling and underlining have been added.

Read it carefully several times and then answer the questions below.

> 1 The sun's rays heat the Earth. However, the space between the sun and the Earth does not contain matter, so heat energy cannot be transferred by the processes of conduction or convection. Instead, the sun transfers heat energy by the process of radiation.
>
> 2 All objects transfer some heat by radiation. The hotter the object, the more heat it radiates. The radiation itself is not hot, but when it is absorbed by an object (it) causes the particles in the object to move more rapidly, thus heating it.

1 Suggest a heading for this extract—one which sums up the main ideas.

____

2 What does it in paragraph 2 refer to? ______________________________

3 What does (it) in paragraph 2 refer to? ______________________________

4 Find a contrast linking word in paragraph 1 ______________ and another in paragraph 2 ______________ .

5 Look at the last sentence in paragraph 1. Note the word instead. What is the author stating that the process of radiation is instead of? ______________________________

____

Exercise 5 Understanding diagrams

Study the two-part diagram on the right then answer these questions.

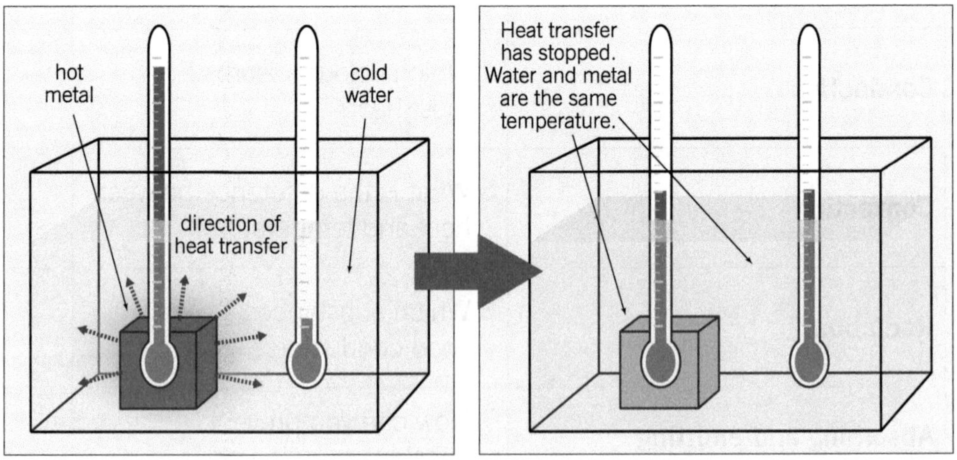

1 What does this diagram show?

____

____

____

____

2 What does the large horizontal arrow represent?

____

____

3 Suggest a caption which summarises what this diagram is showing you.

____

____

Exercise 6 Matching terms and meanings

Match the terms with their meanings, by writing the correct number in each box.

1 radiation	☐	an object that conducts heat well
2 heat	☐	this type of heat transfer can occur in space
3 temperature	☐	a form of energy
4 conductor	☐	the process in which heat energy is transferred by the movement of particles in a gas or a liquid
5 insulator	☐	a measure of how hot or cold something is
6 convection	☐	an object that does not transfer energy well
7 conduction	☐	the process in which energy is transferred from particle to particle in a solid

Exercise 7 Making and justifying decisions

Circle the correct answer to each of the following statements.

1 Heat travels from cold objects to hot objects. TRUE FALSE

2 Heat is measured in joules, temperature is measured in degrees Celsius. TRUE FALSE

3 The heat needed to raise the temperature of 20 g of steel by 1°C is the same as that needed to raise 10 g of steel by 1°C. TRUE FALSE

4 Heat is given off more readily from a black teapot than from a silver teapot. TRUE FALSE

Justify your answer to two of your answers above.

1 ______________________________

____

____

2 ______________________________

____

____

Exercise 8 Writing in role

Imagine you are a science reporter on a magazine. Write a short response (no more than 100 words) to each of the following questions from readers.

Dear Scientist,
I left a shovel out in the hot sun the other day, but the handle was in the shade.
When I went to pick up the shovel, even the handle was hot. How come?
Shoveller

Dear Shoveller,

Dear Scientist,
I am eight years old and I have a pet bird. When the weather is really cold it fluffs up its feathers.
Why does it do this?
Bird Boy

Dear Bird Boy,

Dear Scientist,
I lit our wood fire in the lounge room last night and after a while the whole room was warm. How does the fire heat the whole room even though it doesn't have a fan or anything?
Puzzled

Dear Puzzled,

Roundup

Exercise 9 Writing instructions

Heat is a form of energy and we need to conserve energy for the future of our planet, as well as to save money on gas and electricity bills.

Imagine you are an expert on the conservation of heat energy. You have been given the task of designing and writing a leaflet titled '8 hot tips to save energy in your home' that will be distributed to every household in Australia. Use the information from this chapter to write the tips in the form of instructions. Remember, instructions always start with an action word (verb), e.g. turn ____, avoid ____, use ____, shut ____, cover ____, insulate ____, buy ____ .

8 hot tips to save energy in your home

Exercise 10 Graphing

Ann cannot work out why she feels hotter on some days than on others, even though the air temperature is the same. Peter suggests that it may have something to do with the colour of her clothing. They decide to do an experiment to test this idea. Ann places a thermometer underneath various coloured cloths of the same fabric and puts them in the sun for an hour. She then records the difference between the air temperature and the temperature on the thermometer under the cloth. Here are the results.

white 0.5°C	dark green 3.2°C	navy 3.8°C	pink 1.7°C
black 4.2°C	purple 3.5°C	mauve 2.7°C	pale blue 1.1°C
pale yellow 0.9°C	grey 2.2°C	brown 3.4°C	

Devise a graph to plot Ann's results. **Hint:** arrange the horizontal axis from the lightest colour (white) to the darkest colour (black). Make sure you label the axes.

When you have plotted the results, make a generalisation about colour and temperature.

Your generalisation.

Chapter 8

Body systems

Overview

tendons

ligaments

MUSCLES

muscle contraction

FOOD
carbohydrates
proteins
fats
vitamins & minerals

BLOOD SYSTEM
heart, arteries, veins,
capillaries

DIGESTION
with the help of enzymes

LUNGS
oxygen in
carbon dioxide out

WASTES
carbon dioxide, urine,
faeces

Exercise 1 Working in groups

In this chapter, you will be doing all the activities and exercises in small groups. Working in groups is a very practical and cooperative way of generating ideas and using language to learn. Before commencing the exercises, you need to understand what makes an effective group, and the roles and responsibilities of group members.

1 Read through the lists of characteristics of effective and ineffective groups below. They have been organised according to the behaviour of their members.

2 Circle the three characteristics of effective groups that you think are the most important for successful group work in your classroom.

3 Now circle the three characteristics of ineffective groups that you personally need to keep in mind when you work in a group.

Characteristics of effective and ineffective groups

Effective groups

1 The atmosphere tends to be informal and comfortable, with group members sitting so that all group members are facing each other.

2 There is a lot of discussion in which everyone takes part. Everyone keeps to the point.

3 Everyone understands the task they have to do.

4 The group members listen to each other. Every idea is given a hearing.

5 When there is disagreement, the group is comfortable with this and works towards sorting it out. Nobody feels unhappy with decisions made.

6 People feel free to criticise and say honestly what they think. Group members show that a contribution is valued by using encouraging body language such as nodding, smiling and eye contact; and by giving credit and using first names of people who contribute good ideas.

7 Everyone knows how the other members of the group feel about what is being discussed.

8 When action needs to be taken, everyone is clear about what has to be done, and they help each other.

9 Different people take over the role of leader from time to time.

10 The group knows how well it is working and what is interfering with its progress. It can look after itself.

Ineffective groups

1 The atmosphere reflects indifference or boredom, with little eye contact between group members.

2 Only one or two people talk. Little effort is made to keep to the point of the discussion.

3 Group members are unsure of the task.

4 People do not listen to each other. Some people do not put their ideas forward to the group.

5 Disagreements are not dealt with effectively. They are put to the vote without being discussed thoroughly. Some people are unhappy with the decisions.

6 People are not open about what they are thinking as they fear their ideas will not be valued by others.

7 One or two people are dominant. They ignore other members' feelings.

8 Nobody takes any interest in what has to be done, and nobody offers to help others.

9 Only one or two people make the decisions and act as group leaders.

10 The group does not talk about how it is working or about the problems it is facing. It needs someone to look after it.

What do you know already?

Exercise 2 Discussing

Form groups of four to six, preferably containing students with different learning styles, to discuss and report on the following topics. (See Understanding your preferred learning style on page 18 of this workbook.) Report your discussion decisions to the rest of the class.

1 List your group's five favourite foods.

2 List your group's five most revolting foods.

3 List five things you can do to remain healthy and prevent yourself from being overweight.

4 Which activity can you do for 30 minutes that would use up the most energy?

5 Why do we need food? What does it actually do for our bodies?

6 How does our body break up large pieces of food into pieces small enough to travel to cells in all parts of our body?

Working space

During reading

Exercise 3 Matching words and meanings

After reading pages 178–180

- The process that causes muscles to become shorter and fatter
- Description of muscles that have regular exercise
- Strong fibres which hold bones and joints together
- Strong hard fibres that attach muscles to bones
- Soft, tough ____________ that covers bones and joints and prevents bones rubbing on each other

muscle contraction
tendons
ligament
cartilage
muscle tone

Exercise 4 Writing a paragraph

After reading pages 178–180

In execise 10 on page 8 of this workbook you learned how to write a paragraph. Remember that a paragraph is a parcel of information, usually containing a topic sentence and supporting sentences. The topic sentence states the main idea of the paragraph, and the remaining sentences give details supporting the topic sentence.

This is the structure for the paragraph.

Topic sentence + **Sentences with supporting details**

Use the paragraph on page 8 as a model to write a paragraph for the following task.

Write a paragraph to explain how the biceps and triceps muscles and the tendons work together to cause the lower arm to move.

Topic sentence The biceps and triceps muscles and the tendons work together to cause the lower arm to move. ____________

Supporting sentences ____________

____

____

____

____

____

Exercise 5 Classifying

After reading Food types on page 182

1 Place the foods below into the table of the four main food types. (Some foods will need to be placed into more than one column.)

cheese	fish	eggs	tomatoes	potatoes	milk	liver	grapes
sausages	lemons	soy milk	cabbage	prawns	nuts	bread	parsley
bacon	pumpkin	cake	olives	yoghurt	capsicum	lettuce	apples
butter	oats	spaghetti	honey	chocolate	margarine		

Proteins	Carbohydrates	Fats	Vitamins & Minerals

2 Writing Definitions

In Chapter 6 on page 43 of this workbook, you learnt the pattern for writing definitions. Use this pattern to write definitions for the following.

Proteins ____________________

Carbohydrates ____________________

Fats ____________________

Vitamins ____________________

Exercise 6 Using sequence linking words

Read the following sentence.

> I heard a strange noise then I jumped out of bed and then grabbed my tennis racquet and then I opened the front door and then I crept into the deck and then my brother jumped out from behind a bush to scare me and then he laughed and then I hit him with the racquet and then I went back to bed.

You could make this information more interesting by using a variety of sequence linking words (and shorter sentences!), as in the example below.

> Initially I heard a strange voice and then jumped out of bed, grabbing my tennis racquet. After I opened the front door, I gradually crept onto the deck. Suddenly my brother jumped from behind a bush! When he laughed, I subsequently hit him with my racquet. Finally I went to bed.

Now it's your turn. Write three sentences, each containing two sequence linking words.

Exercise 7 Writing in role

After reading pages 185–186

One way to help yourself to understand abstract ideas is to create stories where you write in the role of the substance to which the sequence of actions is happening. For example, you may have seen a film titled *Inner Space*, a fantasy about a person who is reduced to the size of a blood cell and travels around the human body.

Here is a simple flow diagram of the digestive process.

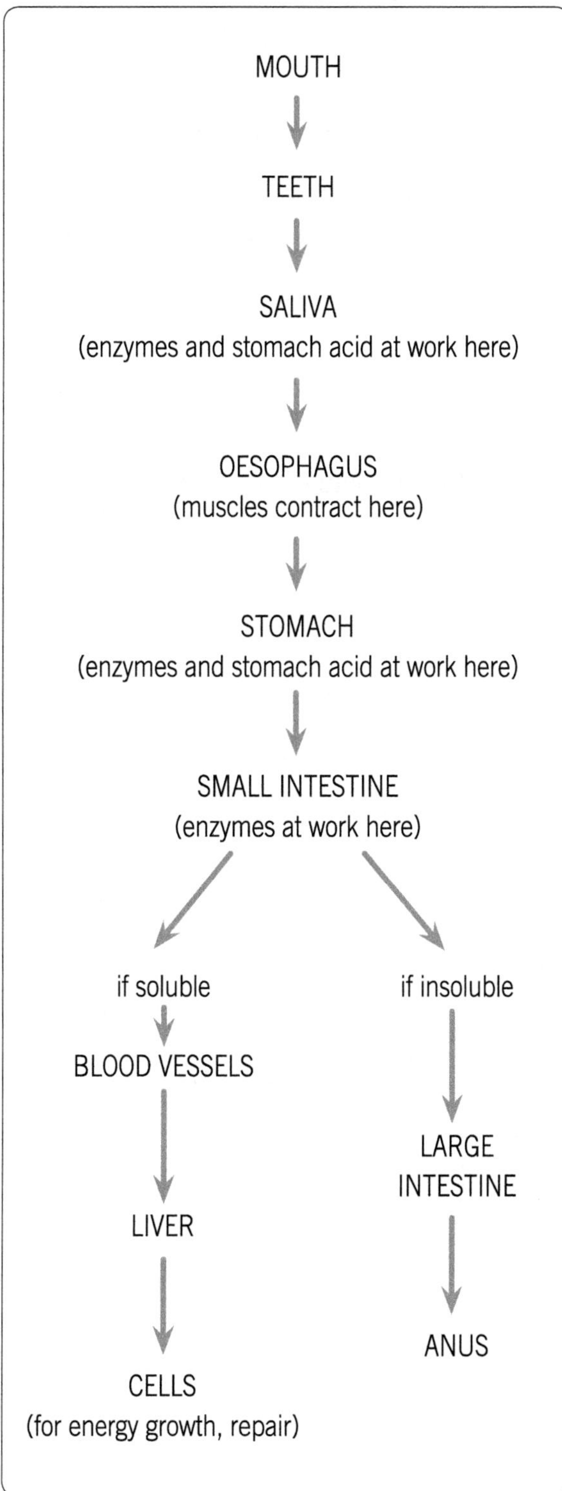

Imagine you are a fried potato chip. You contain starch and oil. Using the flow chart to assist you, describe your journey, your condition, and what is happening to you at each location from beginning to end. Because these events happen to you in a sequence, you should use some *sequence linking words* from page 92 of this workbook.

I am a delicious fried potato chip. I know that some people think that I am fattening because potatoes are a vegetable in the ____________ food group, and oil is one of the ____________ food group, but you need a balanced diet for energy, for growth and to keep ____________________

____

____.

Right now, I am inside a mouth where I am being

____

____

____

____

____

____

____

____

____

____

____

____

____

____

____

____

Exercise 8 Interpreting illustrations

After reading The heart and blood vessels on page 189

1 Examine the diagram on the bottom of page 189 carefully and then answer these questions.

a Which colour is used to illustrate veins? ______________________

b Which colour is used to illustrate arteries? ______________________

c What is represented by the black box on the person's wrist? ______________________

____

2 Examine Fig 6 carefully and then complete this compare and contrast sentence.

Both arteries and veins have ______________________

but ______________________

____

Exercise 9 Comparing and contrasting

After reading How the heart works on page 191

The explanation in the textbook uses the author's words. Use your own words to complete the cartoon bubble below:

Exercise 10 Summarising in a table

After reading pages 193–195

Complete the table below, summarising how your body gets rid of wastes. Gaseous wastes have been completed for you as a model.

	Gaseous wastes	Liquid wastes	Solid wastes
Waste	carbon dioxide	urine	faeces
Composition	carbon dioxide and other gases		
Parts of the body through which the wastes pass	nasal cavity ↓ trachea ↓ bronchi ↓ alveoli ↓ blood ↓ alveoli ↓ nasal cavity	↓ bladder	
Name of the process	respiration	excretion	defecation

Roundup

Exercise 11 Writing in role

Imagine you are a doctor. Write a short response (no more than 100 words) to the following questions in a teenage magazine.

Dear Doctor,
When I was playing netball I fell awkwardly and the coach said I had snapped my Achilles tendon. What is he talking about?
Sally

Dear Sally,

Dear Doctor,
Without my daily chocolate treat, life is just not worth living! Is there any way I can eat chocolate every day and still remain slim, fit and healthy?
Chocoholic

Dear Chocoholic,

Dear Doctor,
What is the difference between the liver and the kidneys? What do they do for us?
Tom

Dear Tom,

Chapter 9

Elements and compounds

Overview

MATTER

Atom → **Molecule**

A molecule is several atoms joined together (atoms may be the same or different).

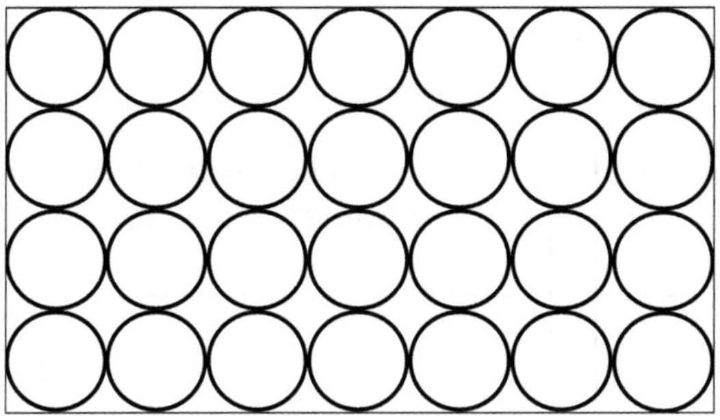

Element

An element is a substance made up of many atoms of the same type. It cannot be broken down to simpler substances.

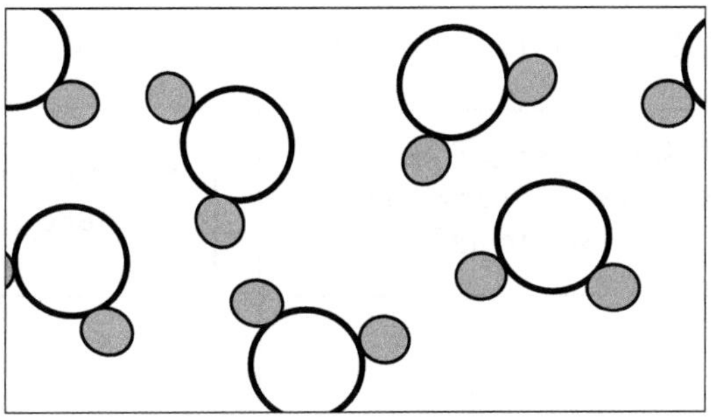

Compound

A compound is a substance made up of two or more different elements chemically combined. It can be broken down into its elements by chemical reactions.

chemical symbol e.g. C, H, O, Ca ↔ chemical formula e.g. H_2O, NaCl

chemical reactions:
- elements → compounds
- compounds → elements

What do you know already?

Exercise 1 Listing in sequences

Using the numbers 1 to 6, mark the following from smallest to largest.

- ☐ DNA molecule
- ☐ hydrogen atom
- ☐ living thing
- ☐ water molecule
- ☐ copper sulfate molecule
- ☐ cell

(After you finish this chapter, you will come back to review your answer.)

During reading

Exercise 2 Explaining

After reading Atoms and molecules on page 202

Sketch the faces and use your own words to answer the question in each box.

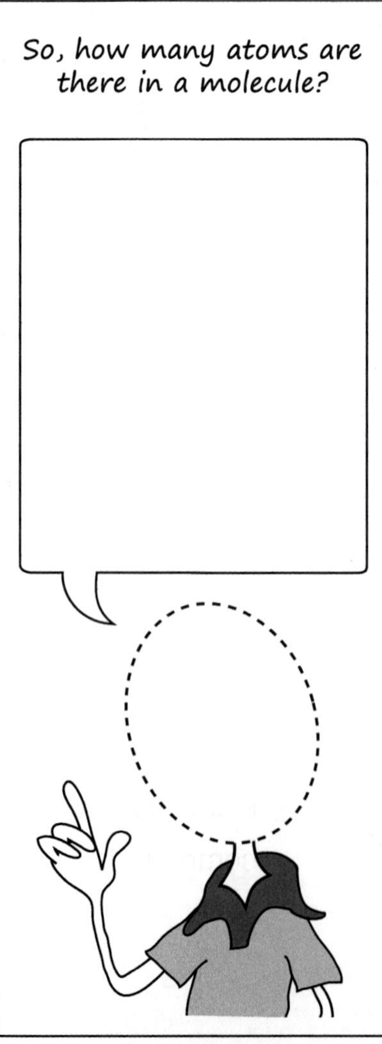

Exercise 3 Making and justifying decisions

After reading Elements on page 204

Decide whether each of the following statements is true or false. Circle your decision and be prepared to justify (give reasons for) your answer.

1 An element contains millions of atoms that are all the same.	TRUE	FALSE
2 Water is an element.	TRUE	FALSE
3 Elements are natural substances.	TRUE	FALSE
4 There are no more elements to be discovered or made.	TRUE	FALSE
5 Gold is expensive because it is a rare element.	TRUE	FALSE

Exercise 4 Interpreting a data table

After reading Elements on page 204 and the Activity on page 206

Use the table on page 206 and the periodic table on page 207 to identify the mystery elements.

1 This element has the symbol Pb. ____________________

2 What is the animal made up of the letters in the symbols for sulphur, hydrogen, argon and potassium?

____

3 What is a girl's name made of the letters in the symbols for aluminium, iodine, sulphur, oxygen and nitrogen? ____________________

4 This element does not melt unless heated to 3500°C. ____________________

5 These elements were discovered four years after Captain Cook sighted Australia. ____________________

6 This element was discovered in 1700. ____________________

7 These were the only two non-metals known in ancient times. ____________________

8 This element is extremely dense. ____________________

9 This metal discovered in the twentieth century is produced in nuclear reactors. ____________________

10 To change this gas into a liquid, you have to cool it to below –200°C. ____________________

11 This is the only metal that is a liquid at room temperature
(i.e. melting point below 20°C and boiling point above 20°C). ____________________

12 This gaseous element is denser than other common gases. ____________________

13 Make up some more clues for elements to try out on your classmates.

Clue 1 ____

Clue 2 ____

Clue 3 ____

Clue 4 ____

Exercise 5 Clarifying ideas and concepts

After reading Compounds on page 210

Complete the explanations.

Exercise 6 Clarifying ideas and concepts

After reading Living and non-living on pages 211–212

Complete the explanation in the cartoon below. There is no right answer … just *your* answer.

Exercise 7 Examining and evaluating

After reading Living and non-living on pages 211–212

1 Think about the following statement from the textbook.

> 'Someday scientists may be able to create life itself.'

2 List three advantages of scientists being able to create life from elements and compounds.

1 ______________________________

2 ______________________________

3 ______________________________

3 List three disadvantages of scientists being able to create life from elements and compounds.

1 ______________________________

2 ______________________________

3 ______________________________

Exercise 8 Completing equations

After reading Chemical reactions and doing Investigations 24 and 25 on pages 214–216

Complete the following chemical equations.

1 sugar $\xrightarrow{\text{heat}}$ [ ] + carbon

2 carbon + oxygen $\xrightarrow{\text{heat}}$ [ ]

3 [ ] + [ ] $\xrightarrow{\text{heat}}$ iron sulfide

4 hydrogen + [ ] $\longrightarrow$ water

Challenge question:

5 sodium + chlorine $\longrightarrow$ [ ]

Roundup

Exercise 9 Labelling

Label each of the following diagrams as either an element, a compound or a mixture.

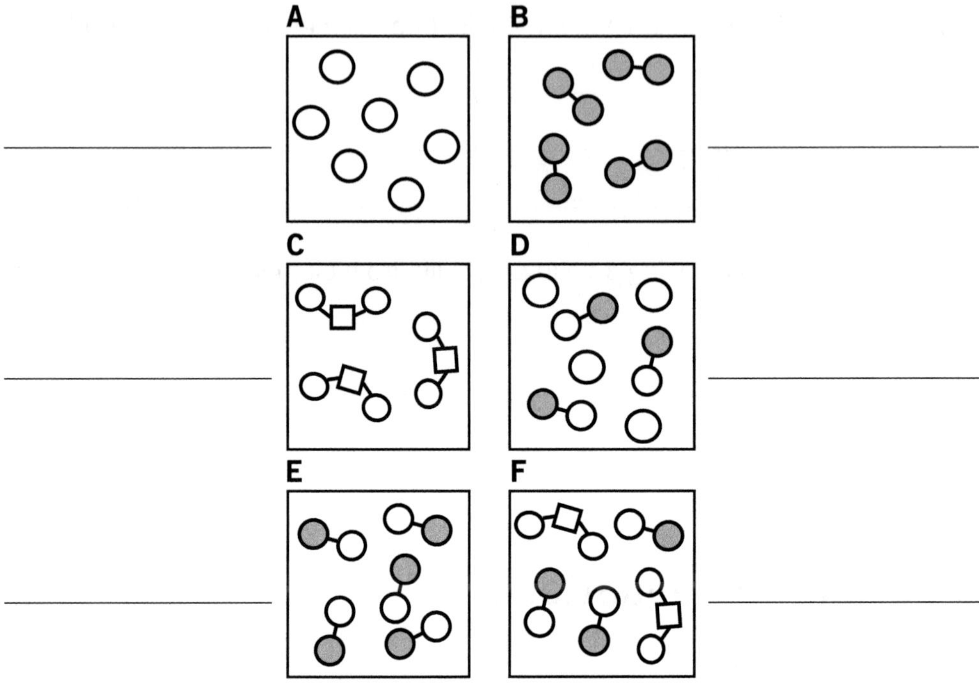

Exercise 10 Expressing a point of view

In exercise 7 you listed arguments for and against scientists creating life. Review your advantages and disadvantages, then decide whether you are for or against scientists creating life. Organise your argument using the following structure. (Use your notebook if you need more space.)

Your opinion—for or against

Reason 1 with supporting statements

Reason 2 with supporting statements

Reason 3 with supporting statements

Conclusion—summing up your argument

Exercise 11 Recalling information

Playing chemical bingo

As a class activity make up small bingo cards (about 10 cm x 10 cm), for example from old manila folders. Each person will need five cards, so for a class of 30 you will need 150 cards. Using the table on page 207 (periodic table), write the symbol for an element on each card. There should be the same number of cards for each element.

Shuffle the cards and deal each person five cards. Place the cards in front of you, face up. Your teacher will call out to the class the names of elements. If you have a card with the symbol of that element, turn the card over. When you have heard the names of all five of your elements, you say 'BINGO!' and you win.

Exercise 12 Reviewing your decisions

Turn back to page 70 of this workbook. Now that you have studied Chapter 9, review your list of building blocks of matter from smallest to largest. Is your sequence correct? If not, correct it.

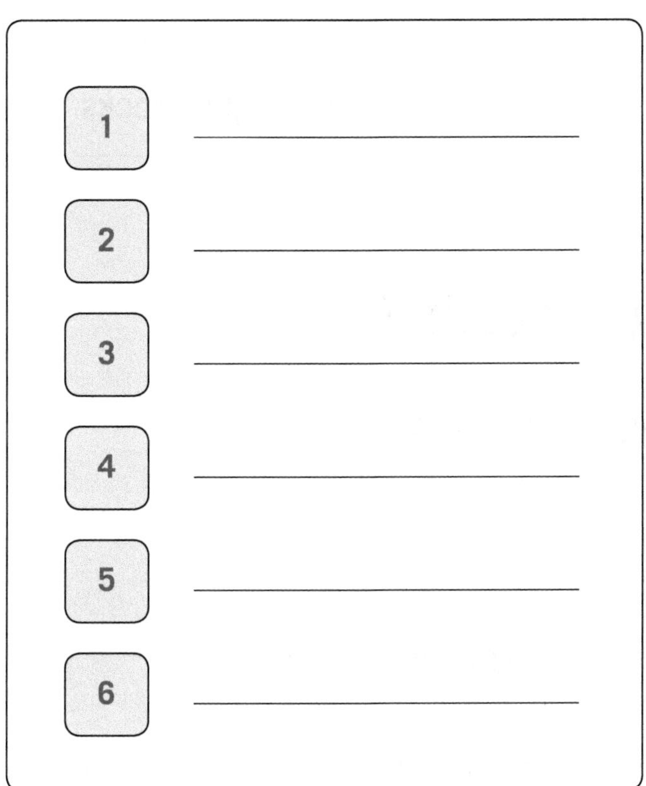

Chapter 10
Rocks

Overview

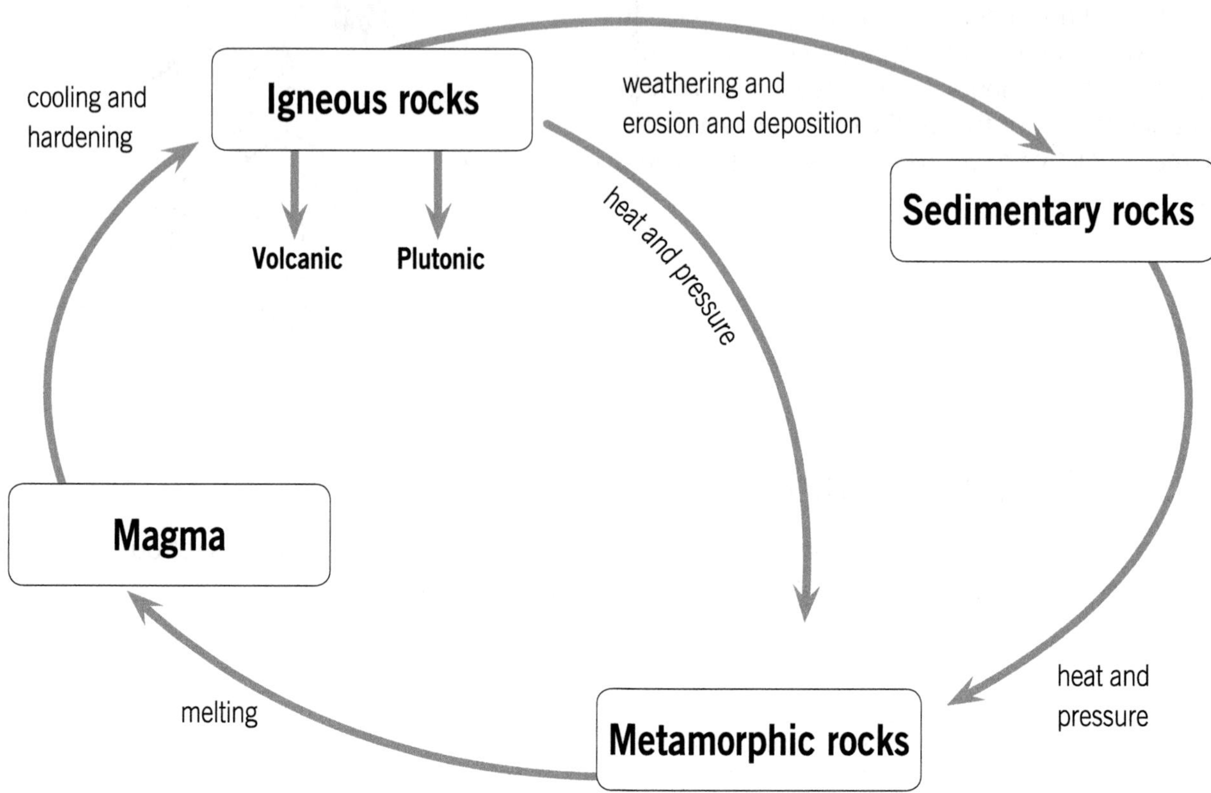

What do you know already?

Exercise 1 Discussing in groups

Discuss the following topics in small groups.

1 What is the inside of the Earth like?

2 What causes volcanoes?

3 How was the Grose Canyon formed? (See Fig 11 on page 235.)

4 How are rocks changed into soil?

5 Where do rocks come from? Why are there many different types of rocks? Can you name at least three different types?

6 What are rocks made of?

During reading

An independent journey

In this chapter of the workbook, you will work independently, at your own pace. You will be working your way through the rock cycle. Use the cycle below to plot your progress by colouring/shading as you work through the ten exercises. Start at STAGE 1 and, before going onto STAGE 2, turn to page 80 of this workbook to check your answers. Do not proceed on your journey unless you have all answers correct, or you have corrected your answers and know why they were not correct.

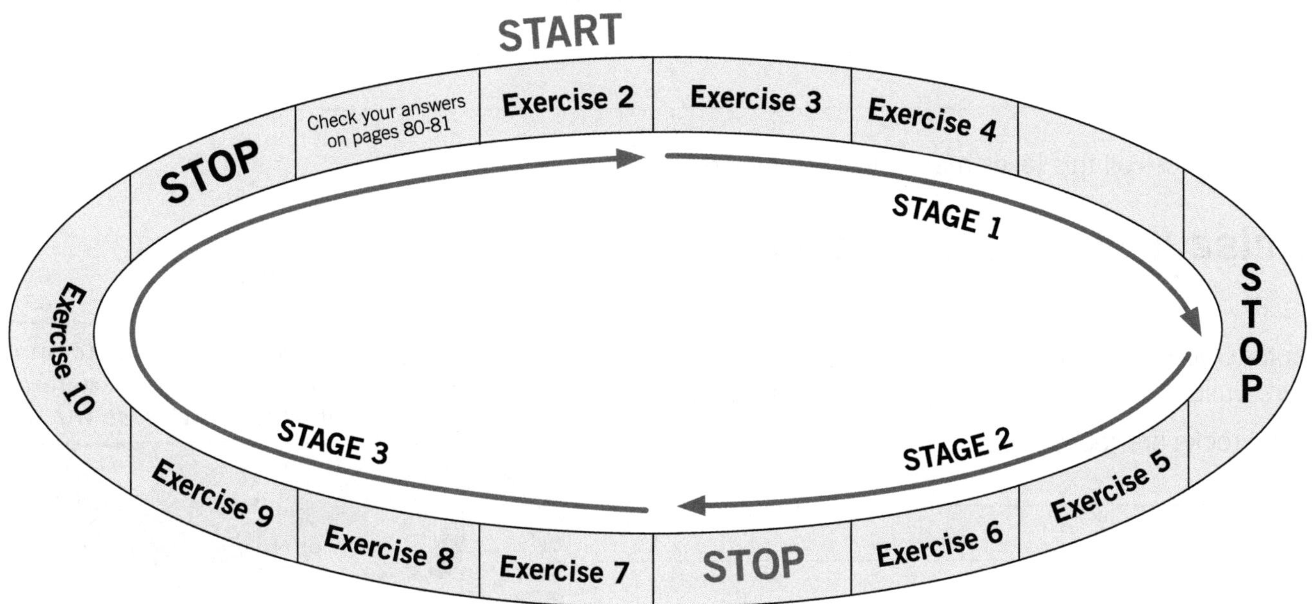

STAGE 1 ROCKS FROM FIRE

Exercise 2 Labelling

After reading Inside the Earth on page 226

Label the illustration on the right with the following:

- crust
- mantle
- outer core
- inner core
- 30 km
- 3000 km
- 6370 km
- lithosphere
- magma

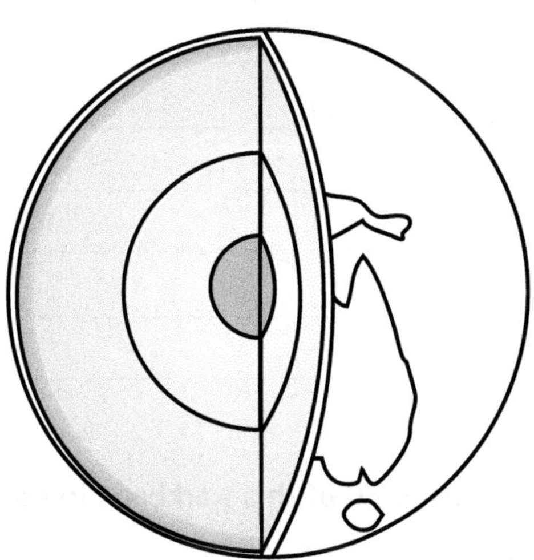

Exercise 3 Explaining in your own words

After reading Movements in the crust on page 226

Complete this dialogue between Sarah and Steve.

Steve: How are mountain ranges formed? They must be pushed up or something.

Sarah: Well, first you need to understand that the crust is ______________________

____

Steve: So how is the crust pushed up?

Sarah: The mantle ______________________

____

____

Steve: I guess all this takes many millions of years.

Exercise 4 Writing definitions

After reading pages 227–228

In Chapter 6 you learnt how to write a definition.
If you are unsure how to do this, go to page 43 of this workbook.

1 Igneous rocks are rocks ______________________

____

____

____

An example is ______________________

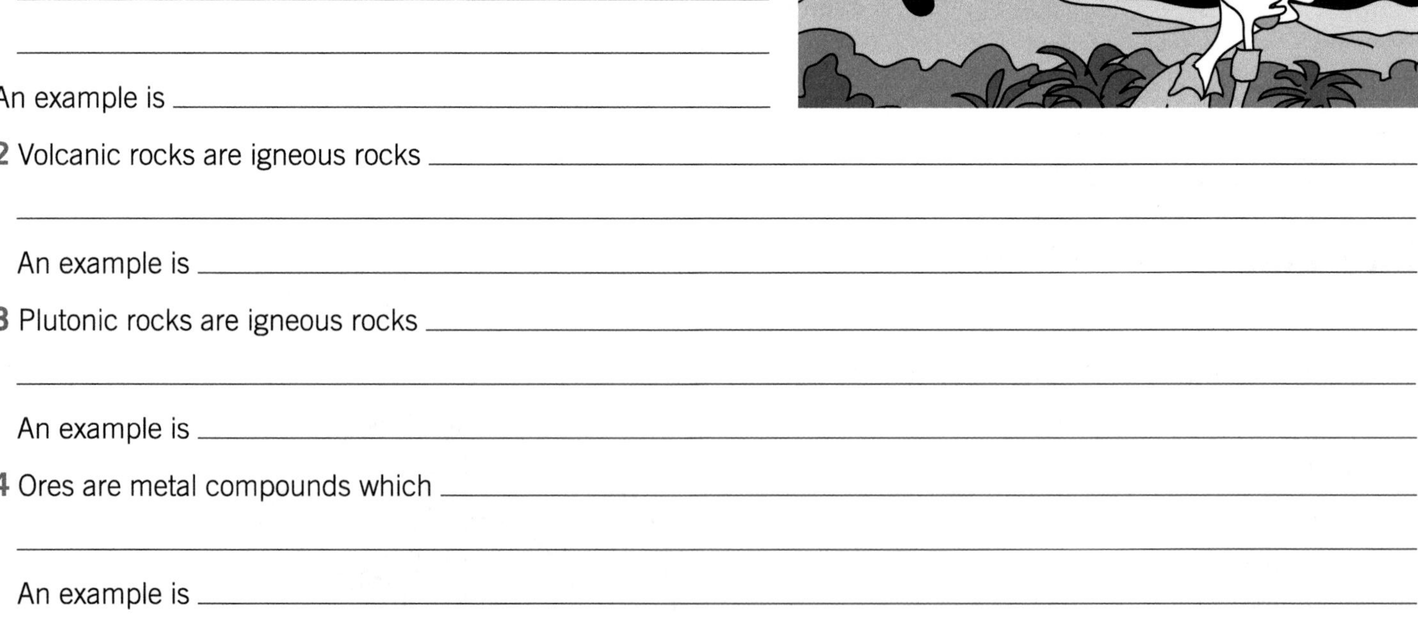

2 Volcanic rocks are igneous rocks ______________________

____

An example is ______________________

3 Plutonic rocks are igneous rocks ______________________

____

An example is ______________________

4 Ores are metal compounds which ______________________

____

An example is ______________________

5 Quartz is a crystal which ______________________

____

Turn to page 80 of this workbook to correct your STAGE 1 answers.

Exercise 5 Completing a diagram summary

After reading pages 232–235

1 Complete the diagram below to summarise the information on weathering.

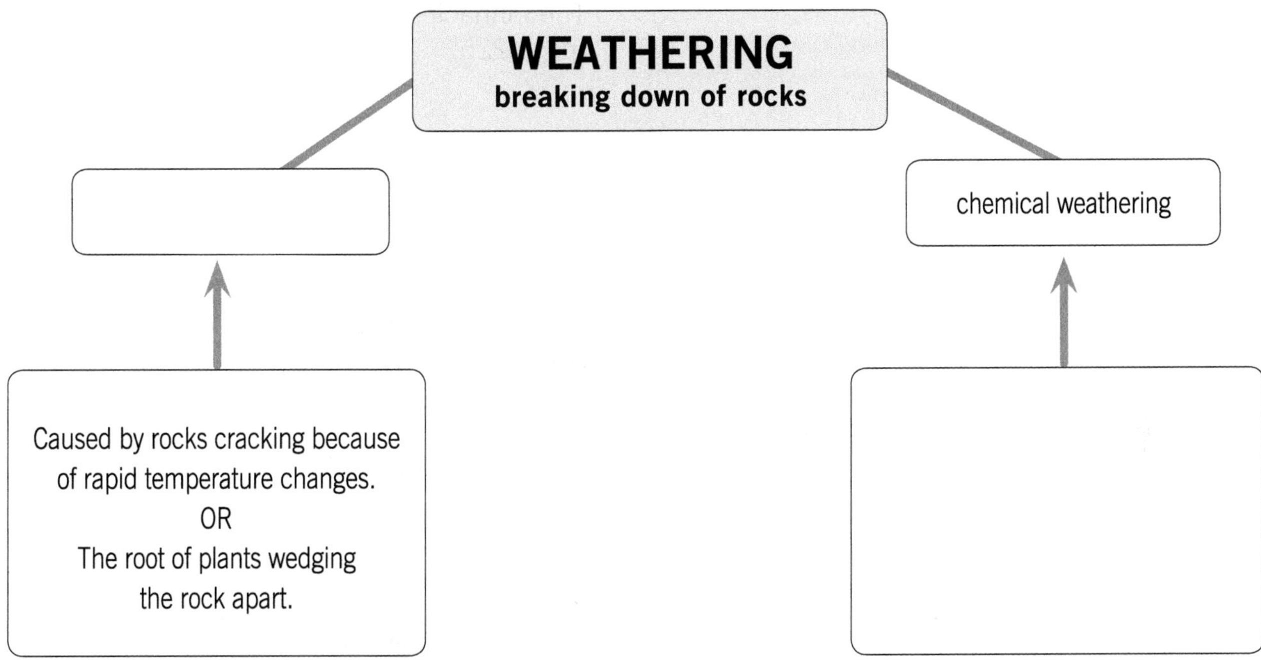

2 Complete the diagram below to summarise the information on erosion.

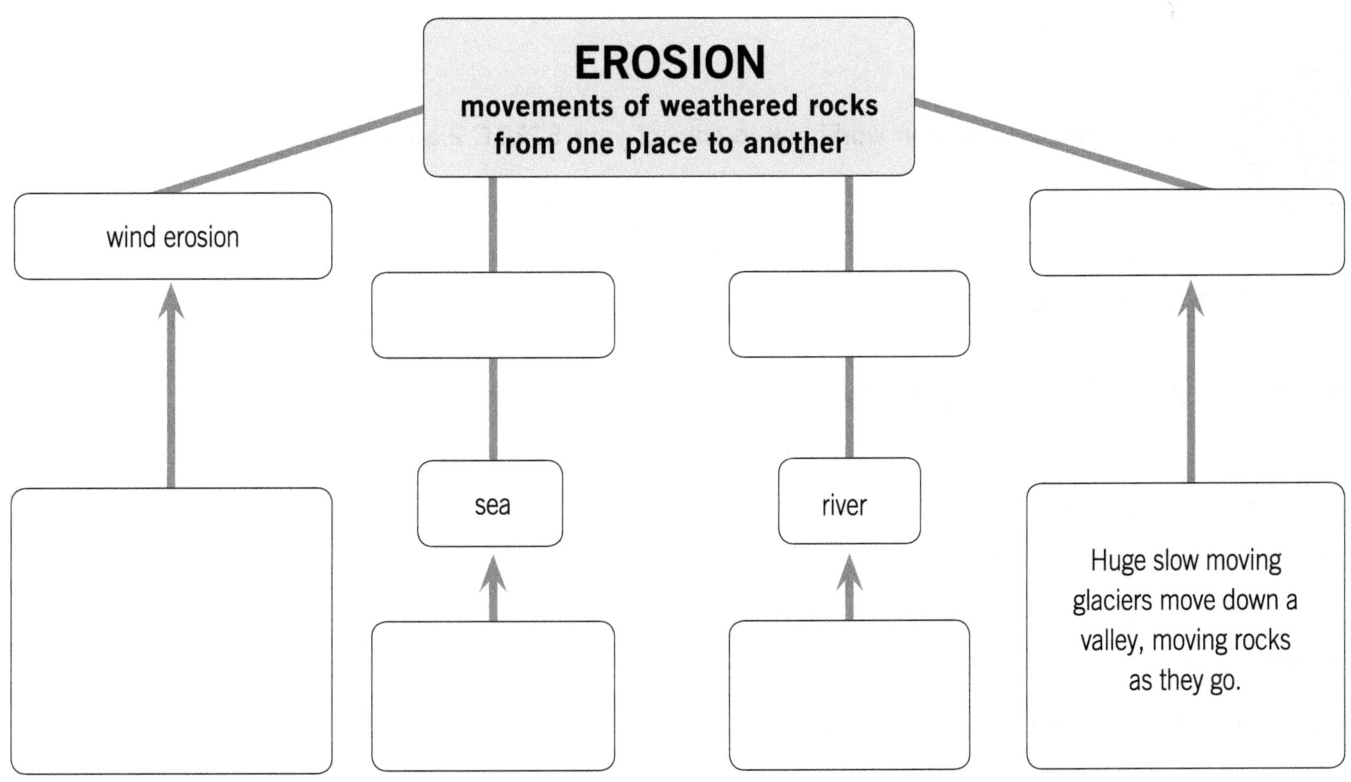

Exercise 6 Linking cause and effect

After reading pages 232–235

Weathering and erosion are the causes of many interesting landforms, illustrated throughout the chapter. In the table below, briefly describe the cause of each effect.

Cause	Effect (Landform)
	The Pinnacles, WA (page 225)
	The Twelve Apostles, Vic (page 224)
	Rounded granite boulders (Fig 6 page 232)
	Limestone caves (Fig 10 page 234)
	The Grose Canyon (Fig 11 page 235)

Turn to page 80 of your workbook to correct your STAGE 2 answers.

STAGE 3 THE ROCK CYCLE

Exercise 7 Writing definitions

After reading Sedimentary rocks on page 238

Complete these definitions.

1 Sedimentary rock is a rock which ______________________________

____

An example is ______________________________

2 Conglomerate is a rock that ______________________________

____

Exercise 8 Inferring

After reading Sedimentary rocks on page 238

Study Fig 15 carefully. Write an inference to explain how those layers of rock formed.

____

____

____

Exercise 9 Completing a table

After reading Metamorphic rocks on page 240

Use the diagram at the bottom of page 240 to complete the table below to show which metamorphic rocks were formed by heat and pressure from the sedimentary rocks.

sedimentary rock	→	metamorphic rock
limestone	→	
sandstone	→	
shale	→	

Exercise 10 Completing a diagram summary

After reading The rock cycle on page 241

Complete the diagram summary by filling in the boxes.

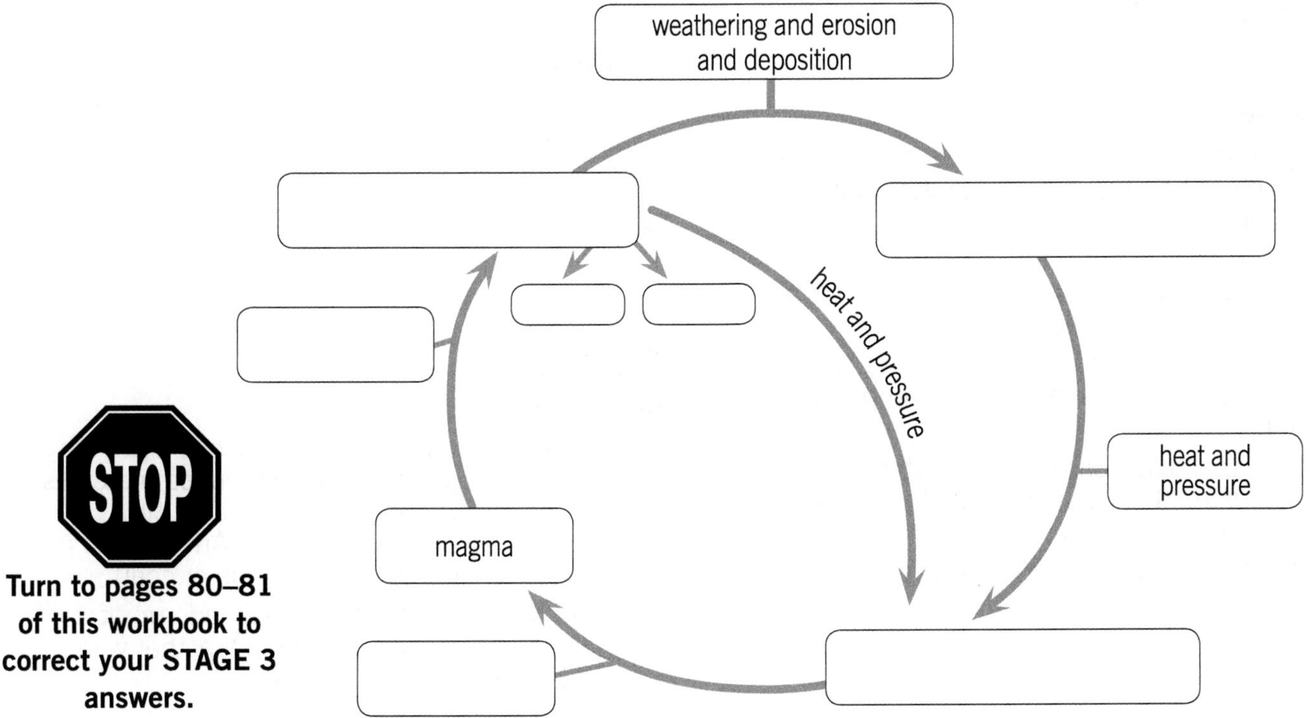

Answers

STAGE 1

Exercise 2

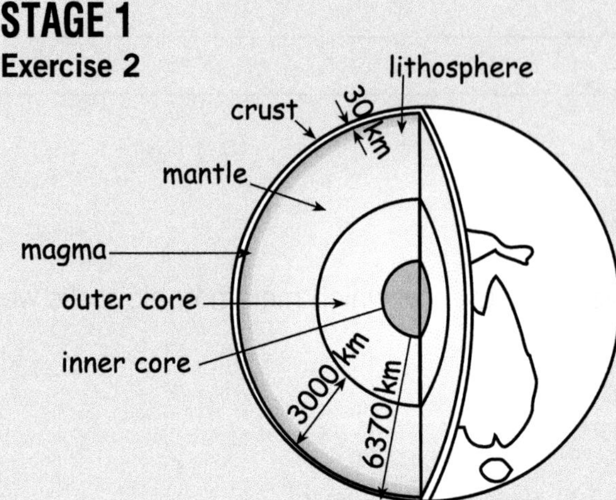

Exercise 3

Steve: How are mountain ranges formed? They must be pushed up or something.

Sarah: Well, first you need to understand that the crust is only very thin and sort of floats on the mantle under it.

Steve: So how is the crust pushed up?

Sarah: The mantle moves all the time and sometimes pushes up the crust. This then slowly turns into rock and where the crust has been pushed up, a mountain range forms.

Steve: I guess all this takes many millions of years.

Exercise 4

1 Igneous rocks are rocks that are formed from the molten rock of the mantle. An example is granite (or basalt).

2 Volcanic rocks are igneous rocks that form when lava cools rapidly on the Earth's surface. An example is basalt.

3 Plutonic rocks are igneous rocks that solidify below the Earth's surface. An example is granite.

4 Ores are metal compounds which can be mined and the metals extracted. An example is bauxite.

5 Quartz is a crystal which has six sides.

STAGE 2

Exercise 5

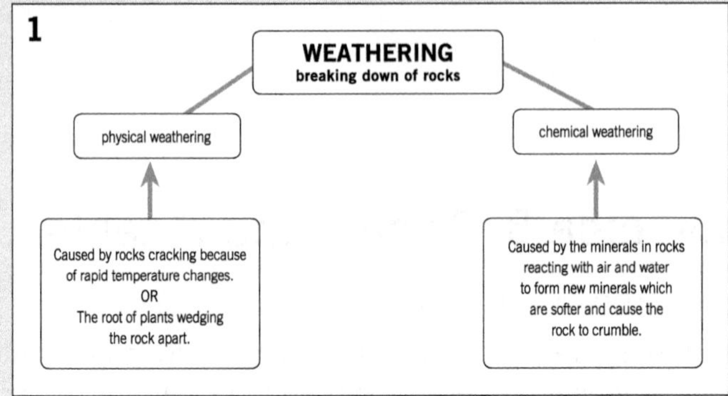

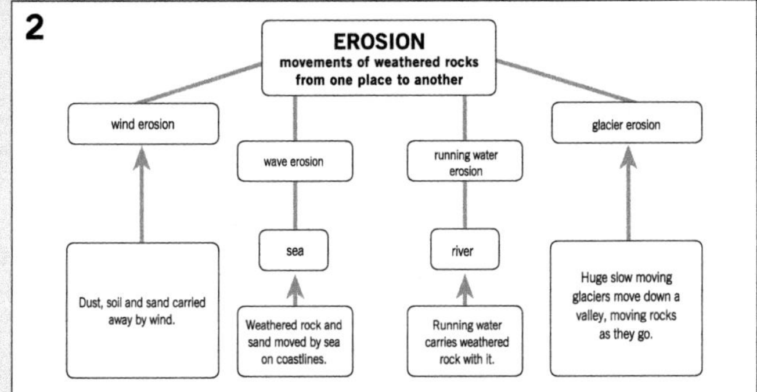

Exercise 6

CAUSE	EFFECT (LANDFORM)
Dust storms wearing away the rocks	The Pinnacles, WA (page 225)
Erosion by the sea	The Twelve Apostles, Vic (page 224)
Chemical weathering in cracks, and physical weathering due to temperature changes	Rounded granite boulders (Fig 14 page 232)
Limestone dissolving in rainwater containing dissolved carbon dioxide	Limestone caves (Fig 19 page 234)
Erosion by river	The Grose Canyon (Fig 20 page 235)

STAGE 3

Exercise 7

1 Sedimentary rock is a rock which is formed by the cementing together of weathered materials such as mud, sand and gravel, and the remains of plants and animals. An example is sandstone (or shale or limestone or coal).

2 Conglomerate is a rock that is made up of pebbles and stones cemented together with sand, clay and other minerals.

Answers

Exercise 8

This is a difficult exercise, but if you have something like this you are on the right track.

Sediments are carried by rivers into seas or lakes. The sediments settle at the bottom and are compressed to form a layer of rock. More sediments are deposited on top of the layer forming other layers. The sedimentary rock layers are pushed up to the surface by the Earth's movements.

Exercise 9

sedimentary rock	→	metamorphic rock
limestone	→	marble
sandstone	→	quartzite
shale	→	slate

Exercise 10

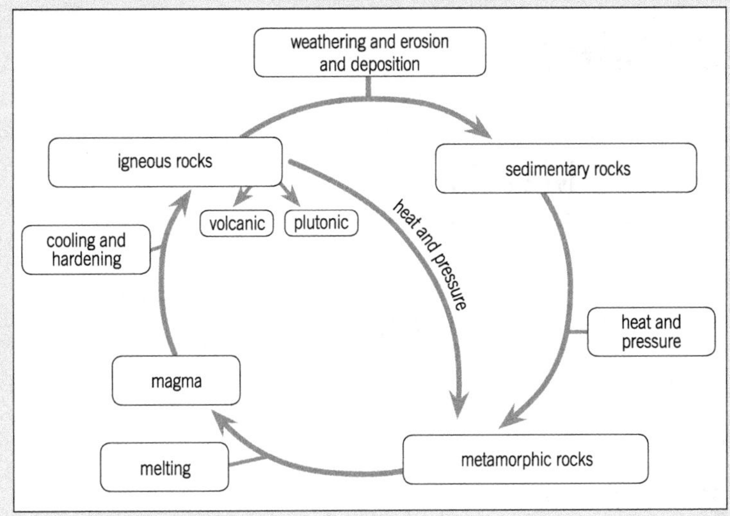

Roundup

Exercise 11 Classifying

Each student in the class is to select a specific rock mentioned in Chapter 10. Use the information to form a 'profile' of the rock so that you can answer questions about it.

Oh Jasper, I think only of you!

Rules:

1 You may only answer 'yes' or 'no'.

2 Other members of the class must ask the questions to obtain a 'yes' or 'no' response. (There are examples below.)

3 If they cannot guess the rock you are thinking of in 6 questions, you're the winner.

Examples of questions the class could ask.

- Are you thinking of an igneous rock? A sedimentary rock? A metamorphic rock?
- Are you thinking of a volcanic rock? A plutonic rock?
- Are you thinking of a sedimentary rock that was formed from other rocks? From the remains of organisms? From the remains of dead plants?

Chapter 11

Everyday substances

Overview

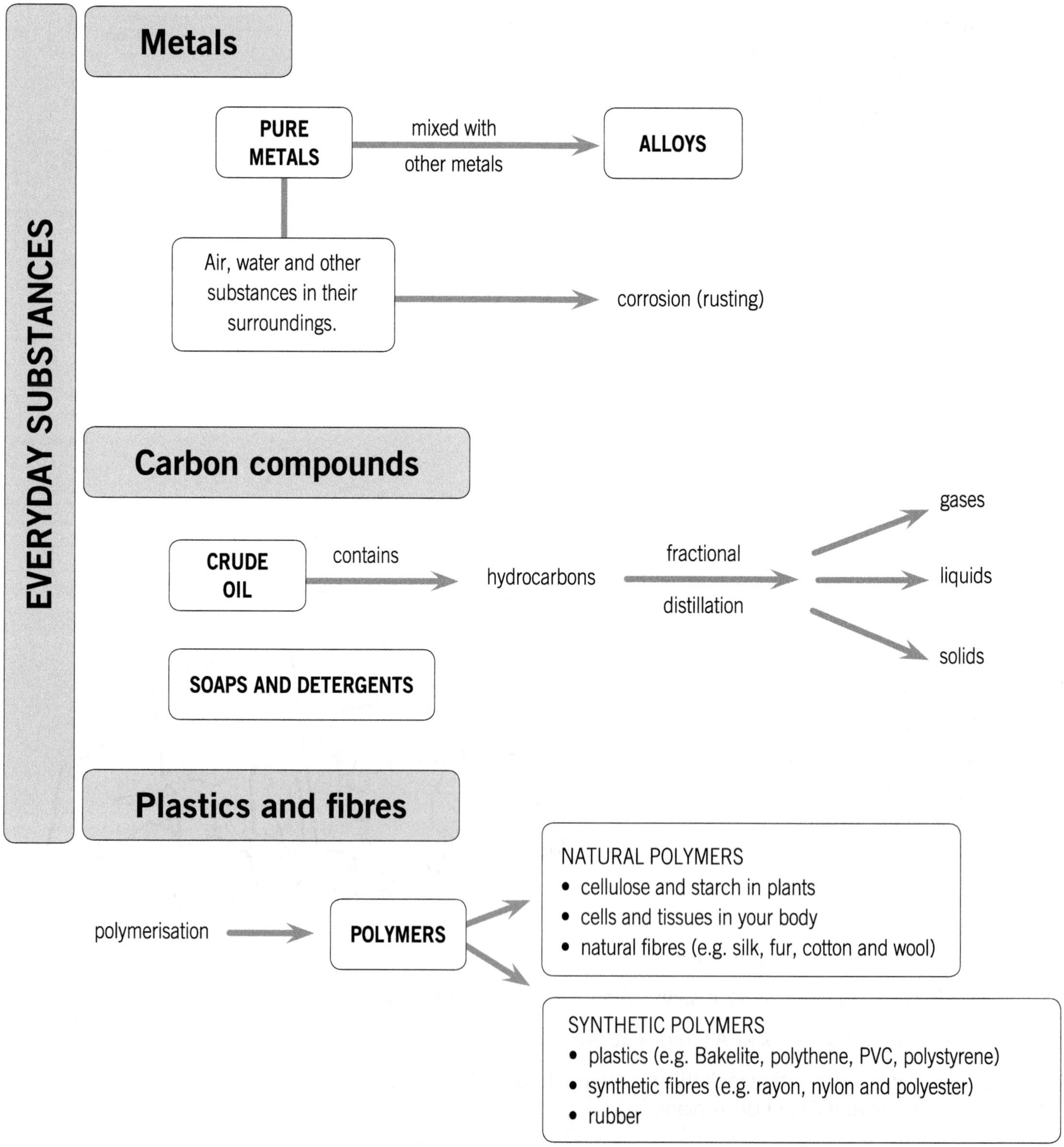

EVERYDAY SUBSTANCES

Metals

PURE METALS → mixed with other metals → ALLOYS

PURE METALS — Air, water and other substances in their surroundings. → corrosion (rusting)

Carbon compounds

CRUDE OIL → contains → hydrocarbons → fractional distillation → gases, liquids, solids

SOAPS AND DETERGENTS

Plastics and fibres

polymerisation → POLYMERS →

NATURAL POLYMERS
- cellulose and starch in plants
- cells and tissues in your body
- natural fibres (e.g. silk, fur, cotton and wool)

SYNTHETIC POLYMERS
- plastics (e.g. Bakelite, polythene, PVC, polystyrene)
- synthetic fibres (e.g. rayon, nylon and polyester)
- rubber

What do you know already?

Exercise 1 Observing and listing

The things around you are made of many different substances. Take a walk around your home or school. List at least 10 different substances you can see that would not have been available to craftsmen 100 years ago.

1 ____________________ 6 ____________________

2 ____________________ 7 ____________________

3 ____________________ 8 ____________________

4 ____________________ 9 ____________________

5 ____________________ 10 ____________________

During reading

Exercise 2 Inferring

After reading Metals on pages 251–252

Suggest why:	Answer	Suggest what metal:	Answer
aircraft parts are not made of pure iron.		aircraft parts are made of.	
musical instruments are not made of Nitinol.		guitar or piano strings are made of.	
dental braces are not made of stainless steel.		dental braces are made of.	
surgical instruments are not made of solder.		surgical instruments are made of.	

Exercise 3 Summarising in a table

After reading Corrosion on page 253

Complete the table below.

Substance/object	Protective process	Why this process prevents corrosion
pure iron		When these metals are added to pure iron, stainless steel which does not rust is formed.
	painting	
bike chains		
	coating with plastic	
steel cans containing food		
cutlery		
	galvanising	
		The zinc bar corrodes instead of the steel around it.

Exercise 4 Finding and applying patterns

After reading Carbon compounds on pages 255–257

1 Examine this list of hydrocarbons.

Methane CH_4 Ethane C_2H_6 Propane C_3H_8 Butane C_4H_{10}

Find the pattern linking the number of carbon atoms and the number of hydrogen atoms.

2 Use this pattern to complete these hydrocarbon formulas.

a $C_5H_?$ ⟶ ____________

b $C_?H_{18}$ ⟶ ____________

c $C_{20}H_?$ ⟶ ____________

d $C_{11}H_?$ ⟶ ____________

e $C_?H_{30}$ ⟶ ____________

f $C_{16}H_?$ ⟶ ____________

Exercise 5 Explaining

After reading Carbon compounds on pages 255–257

Complete the cartoon using your own words.

Exercise 6 Translating from illustration to table

After examining Fig 10 on page 257

Complete the table on the right.

GASES
hydrocarbons with a low boiling point condense here
LIQUIDS
furnace
fractionating column
crude oil
hydrocarbons with a high boiling point condense here
SOLIDS

Names of hydrocarbons	Boiling range (°C)	Strength of inter-molecular forces	Uses

Exercise 7 Comparing and contrasting

After reading Soaps on pages 257–259 and Detergents on page 260

1 Complete the table below.

	Soaps	Detergents
Ingredients		
Synthetic or made from natural products		
Household uses		
Biodegradable or non-biodegradable		
Disadvantages of each product compared with the other		

2 Examine the table above carefully. Look for similarities and differences. Complete the following sentences describing similarities and differences.

a Both soaps and detergents ______________________________

____

____

____

b Soaps and detergents are different in that ______________________________

____

____

____

Exercise 8 Sequencing

After reading page 259

Note: Soaps and detergents remove grease in the same way.

Read the following statements carefully, then mark them in the correct order by placing the numerals 1 2 3 4 5 in front of them.

_______ The negatively charged heads surround the grease particles with a shield of negative charges.

_______ Soap is put into a solution containing a greasy cloth.

_______ As the solution is agitated, the grease surrounded by the soap molecules is lifted off the cloth.

_______ The grease-loving 'tails' stick into the surface of the grease, leaving the 'heads' out in the solution.

_______ A soap molecule has a negatively charged 'head' which loves water, and a hydrocarbon 'tail' which hates water but is attracted to grease.

Exercise 9 Explaining

After reading pages 262–263

1 Complete the cartoon using your own words.

2 Using the information in the blue box on page 263 as a guide, complete the diagram showing how vinyl chloride is polymerised to PVC.

H, Cl: H–C=C–Cl / H H	H, Cl: H–C=C–Cl / H H	H, Cl: H–C=C–Cl / H H	H, Cl: H–C=C–Cl / H H

The double bonds break, leaving spare links.

New single bonds form.

Exercise 10 Comparing and contrasting

After reading Properties and uses of plastics on page 264

1 Complete the table below.

	Thermoplastics	Thermosets
Diagram of structure		
Describe the attractive forces between chains		
Describe the effect of heat on this plastic		
Characteristics of this type of plastic compared to the other type		
Common examples of this type of plastic		
Household uses for these plastics		

2 Use *compare and contrast linking words* from page 92 of this workbook to write sentences to compare and contrast each of the information pairs above. The first one is done for you as an example.

a Both thermoplastics and thermosets are made from polymers. However thermoplastics contain long thin polymer chains which can slide over each other and can be separated by heat, whilst thermosets contain a network of linked chains which have strong bonds between them and which cannot be separated by heat.

Exercise 11 Comparing advantages and disadvantages

After reading Recycling and reusing plastics on page 267

Complete the table below.

Recycling method	Advantages of this recycling method	Disadvantages of this recycling method
melting and remoulding		
		lower quality product
	The system can handle 8000 tonnes of plastic waste in a year. Valuable oil products are produced and the uncondensed gases heat the plastic waste.	

Exercise 12 Defining

After reading Fibres on pages 268–270

Here is an example of how to use the diagram on the bottom of page 268 to define natural fibres.

> Natural fibres are fibres used in fabrics to make clothes that may be made from either plant or animal material. Examples are cotton, linen, wool, silk and fur.

Use the example above and the diagram on page 268 to define synthetic fibres.

Synthetic fibres are

____

____

____

____

Exercise 13 Understanding on three levels

After reading pages 269–270

For each of the statements below, write T (true) or F (false) in the brackets provided. When you have finished, discuss your decisions in small groups. Try to reach agreement on your answers as a group. You should be ready to justify (give reasons for) your answers.

Level 1: Reading for accuracy

(Be able to show where statements in the textbook support your answers.)

1 Each hair on your head contains two sulfur atoms. ( )

2 When straightening solution is combed into hair, many of the sulfur bonds break. ( )

3 Hair is made of protein. ( )

Level 2: Drawing conclusions

(Be able to use information in the textbook to show how you arrived at this conclusion.)

1 S–S bonds are impossible to break. ( )

2 Neutraliser is used to stop the 'rotten egg' smell. ( )

3 Hair is weakened by having it straightened too often. ( )

Level 3: Applying your knowledge

(Be able to use your knowledge in a wider context.)

1 Straight hair would be able to be curled using the straightening method in reverse. ( )

2 Naturally curly hair has more cross-links between molecular chains than naturally straight hair. ( )

3 Hair can be lengthened by pulling it and then applying the straightening solutions. ( )

Roundup

Exercise 14 Recalling information

1 Answer each clue and use the circled letters to make another key word in Chapter 11.

Clue	Letter
Carbon compound made by boiling vegetable or animal oils with caustic soda	P
Another word for rusting	O
Process in which iron is coated with zinc	L
A metal mixture that has different properties from those of the metals it is made from	Y
The main metal in duralium	M
Plastics which do not melt when heated	E
The process in which crude oil is separated	R
The process in which crude oil is separated	I
Brown iron oxide	S
Discoloured—referring to gold, silver and brass	A
Synthetic soap made from crude oil	T
The opposite of natural	I
Plastics which stretch and bend easily and soften on heating	O
The substance which is added to iron to make steel	N

2 Use the 12 circled letters to make another key word in this chapter. (**Hint:** the word begins with H.)

3 Write a clue for this word.

Linking words

Sequence linking words (for actions that occur one after the other)

first, second	then	before	finally	subsequently
to begin with	later	next	gradually	initially
when	from there	last	initial	after

Cause and effect linking words (for actions which cause another action to occur)

therefore	so that	an outcome of	results in
due to	as a result of	causing	consequently
since	because	caused by	this leads to
the effect of	the reason for	if … then	when … then
resulting in	making	as	accounts for
so	thereby	causes	gives rise to
creates	owing to	hence	thus

Contrast linking words (to describe differences)

on the other hand	in contrast to	different from	however
alternatively	but	although	rather than
whereas	while	unlike	yet
even though	less than	more than	whilst

Comparison linking words (to describe similarities)

both	same	in both	like
in both cases	similarly	and	as well as
in the same way	just as … so …	alike	similar in that

Generalisation linking words (for statements that are true in most cases)

generally	usually	commonly	many
normally	as a general rule	most	mostly